Meteorology

Meteorology

Eleanor Flores

SYRAWOOD
PUBLISHING HOUSE
New York

Published by Syrawood Publishing House,
750 Third Avenue, 9th Floor,
New York, NY 10017, USA
www.syrawoodpublishinghouse.com

Meteorology
Eleanor Flores

International Standard Book Number: 978-1-64740-111-5 (Hardback)

Cataloging-in-Publication Data

Meteorology / Eleanor Flores.
p. cm.
Includes bibliographical references and index.
ISBN 978-1-64740-111-5
1. Meteorology. 2. Atmospheric science. 3. Climatology. I. Flores, Eleanor.
QC861.3 .M48 2022
551.5--dc23

TABLE OF CONTENTS

PREFACE

This book has been written, keeping in view that students want more practical information. Thus, my aim has been to make it as comprehensive as possible for the readers. I would like to extend my thanks to my family and co-workers for their knowledge, support and encouragement all along.

The branch of atmospheric sciences that makes use of atmospheric chemistry and physics is known as meteorology. It focuses on weather forecasting. Meteorological studies take place at various scales such as microscale, mesoscale, synoptic scale and global scale. Important meteorological principles include boundary layer meteorology and dynamic meteorology. Boundary layer meteorology studies the processes in the layer of air that is directly above the Earth's surface. Dynamic meteorology is concerned with the fluid dynamics of the atmosphere. Some of the areas where meteorology is applied are aviation, agriculture and military. This book elucidates the concepts and innovative models around prospective developments with respect to meteorology. Such selected concepts that redefine this field have been presented in it. The coherent flow of topics, student-friendly language and extensive use of examples make this book an invaluable source of knowledge.

A brief description of the chapters is provided below for further understanding:

Chapter - Introduction

The branch of atmospheric science which focuses on weather forecasting is referred to as meteorology. It utilizes atmospheric chemistry and atmospheric physics. This is an introductory chapter which will briefly introduce the significant aspects of meteorology.

Chapter - Meteorological Occurrences

The most common weather occurrences which are studied under the domain of meteorology are clouds, rain, water vapor, winds, storms, snow, tropical cyclones, etc. The topics elaborated in this chapter will help in gaining a better perspective about these meteorological occurrences.

Chapter - Branches of Meteorology

Some of the major branches of meteorology are microscale meteorology, mesoscale meteorology, synoptic scale meteorology and hydrometeorology. The chapter closely examines the key concepts of these branches of meteorology to provide an extensive understanding of the subject.

Chapter - Weather Forecasting

The application of science and technology for the prediction of atmospheric conditions of a given location and time is referred to as weather forecasting. Various types of weather forecasting include marine weather forecasting, numerical weather prediction and nowcasting. The diverse applications of these types of weather forecasting have been thoroughly discussed in this chapter.

Chapter - Meteorological Instruments

Various equipment are used in meteorology such as hygrometer, anemometer, barometer, ceiling balloon, radiosonde, SODAR, barograph, wind profiler, ceiling projector, dark adaptor goggles, windsock, etc. The topics elaborated in this chapter will help in gaining a better perspective about these equipments used in meteorology.

Chapter - Applications of Meteorology

Meteorology is applied in various sectors such as aviation, construction and energy sector. It is also used to study areas of disturbed weather in order to recognize potential tropical cyclones. The topics elaborated in this chapter will help in gaining a better perspective about these applications of meteorology.

Eleanor Flores

Introduction

CHAPTER 1

The branch of atmospheric science which focuses on weather forecasting is referred to as meteorology. It utilizes atmospheric chemistry and atmospheric physics. This is an introductory chapter which will briefly introduce the significant aspects of meteorology.

METEROLOGY

Meteorology is the study of the Earth's atmosphere and the variations in temperature and moisture patterns that produce different weather conditions. Some of the major subjects of study are such phenomena as precipitation (rain and snow), thunderstorms, tornadoes, and hurricanes and typhoons.

The importance of meteorological events is felt in various ways. For example, a drought results in water shortages, crop damage, low river flow rates, and increased wildfire potential. In addition, these effects may lead to restricted river travel, saltwater infiltration in aquifers and coastal bays, stress on various plant and animal species, population shifts, economic hardship, and even political unrest. The critical impact of weather on human activity has led to the development of the uncertain science of weather forecasting.

Aristotle's Meteorologica (340 B.C.) concerned all phenomena above the ground. Astronomy, including the study of meteors, or "shooting stars," later became a separate discipline. The science of meteorology was restricted eventually to the study of the atmosphere. Various weather phenomena are still referred to as "meteors," such as hydrometeors (liquid or frozen water — rain, snow and snowflakes, clouds, fog), lithometeors (dry particles — sand, dust, or smoke), photometeors (optical phenomena — halos, mirages, rainbows, coronas), and electrometeors (electrical phenomena — lightning, Saint Elmo's fire).

Modern meteorology focuses primarily on the typical weather patterns observed, including thunderstorms, extratropical cyclones, fronts, hurricanes, typhoons, and various tropical water waves. Meteorology is usually considered to describe and study the physical basis for individual events. In contrast, climatology describes and studies the origin of atmospheric patterns observed over time. Several important phenomena, such as monsoons and the El NiÃ±o–Southern Oscillation, are considered in both meteorology and climatology because they exhibit large changes on seasonal time scales.

Atmosphere and it's Layers

The atmosphere is the blanket of gases which surrounds Earth. It is held near the surface of the planet by Earth's gravitational attraction.

Without the atmosphere there could be no life on Earth. The atmosphere:

- Contains the air we breathe;
- Protects life from harmful radiation from the Sun;
- Helps keep the planet's heat from the Sun from escaping back into space;
- Is a major element of the water cycle;
- Keeps the climate on Earth moderate compared to that of other planets.

The atmosphere is made up of a mixture of gases, mostly nitrogen, oxygen, argon and carbon dioxide. It reaches over 500km above the surface of the planet. There is no exact boundary between the atmosphere and outer space. Atmospheric gases become thinner the higher up you go. The atmosphere just keeps getting less and less dense, until it "blends" into outer space.

The atmosphere is divided into four layers based on temperature: the troposphere, stratosphere, mesosphere, and thermosphere.

Troposphere

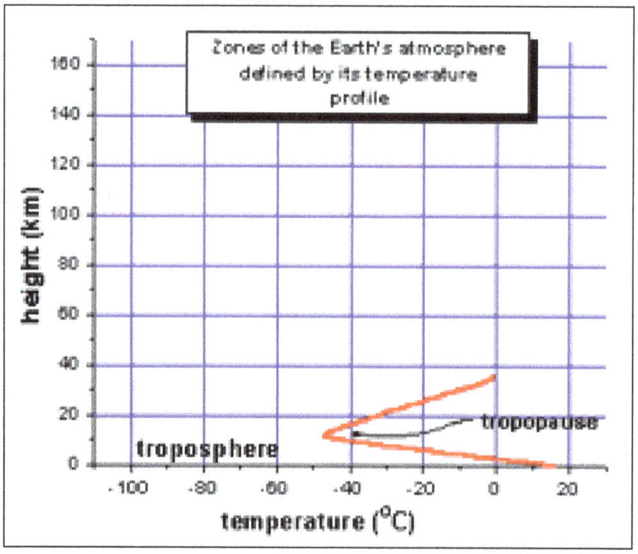

This is the lowest layer of the atmosphere, closest to the surface of the earth. The troposphere varies in height in different parts of the world, from about 8km above sea level at the poles, to 16km at the equator. Within the troposphere, the temperature drops rapidly the higher you go.

This is the layer where we see clouds and most of the "weather" occurs. The top layer of the troposphere is called the tropopause. Within the troposphere the temperature drops to a low of -56°C. This marks the beginning of the tropopause. Through the tropopause, the temperature reverses and begins to increase. The height of the tropopause varies from the poles to the equator, but also from the summer to winter.

Stratosphere

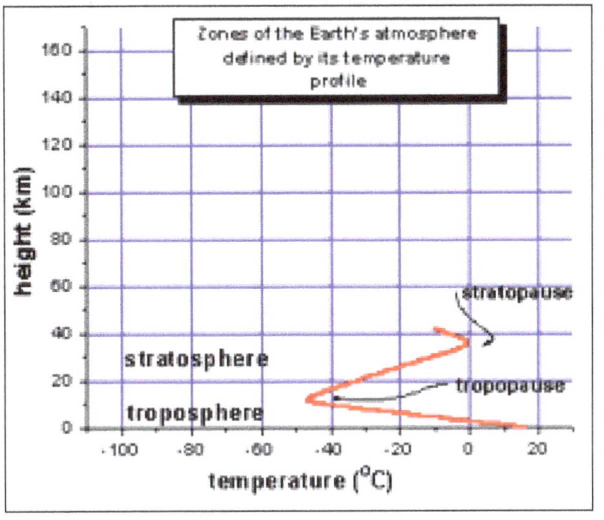

For a distance of about 18km above the tropopause, there is a layer called the stratosphere. In this layer, the pressure continues to decrease, but the temperature increases gradually to 0°C.

Like the troposphere, the stratosphere also varies in thickness. It is quite deep over the poles and thinner over the equator.

The stratosphere contains a layer of ozone which absorbs the Sun's ultraviolet rays, protecting life on the earth's surface.

The top layer of the stratosphere is called the stratopause. In this layer the temperature, once again, begins to fall.

Mesosphere

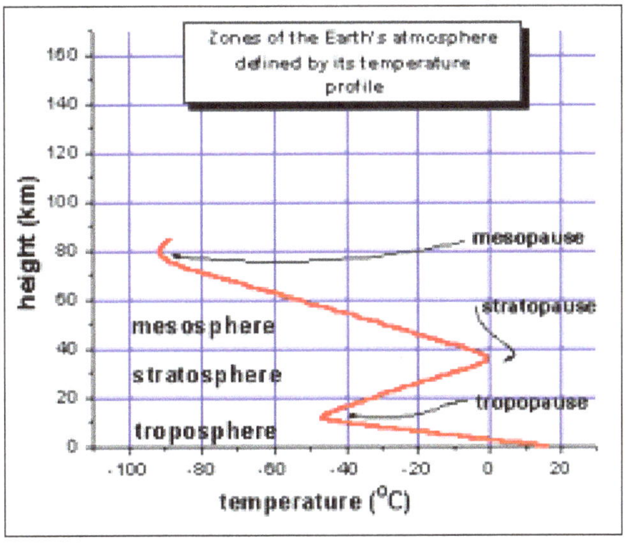

The mesosphere reaches up to about 80km above the surface of the earth. This is the coldest layer of the atmosphere, where the temperature drops rapidly with altitude.

In the top layer of the mesosphere, called the mesopause, the temperature bottoms out at a low of about -100°C at 80km above the earth. After that the temperature begins to rise again with greater altitude.

This is the layer where meteorites usually burn up as they enter the atmosphere.

Thermosphere

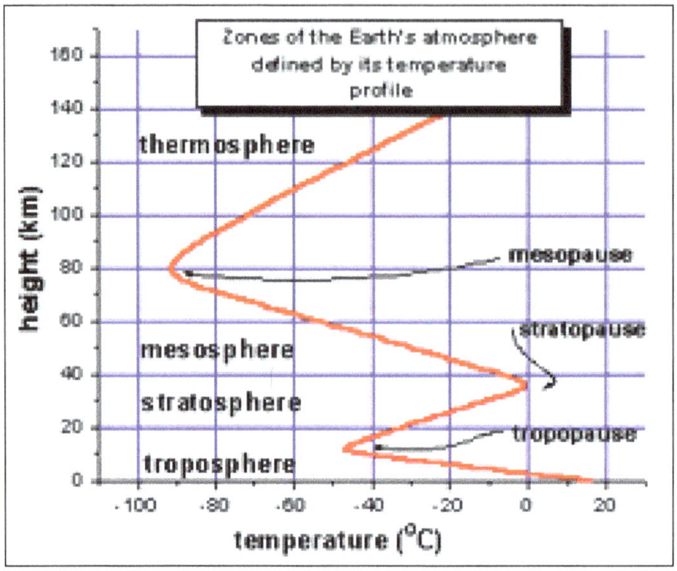

The thermosphere extends upwards from a point 80-100 kilometres above the earth's surface. There is very little air in this layer. The temperature continues to rise in the thermosphere and beyond, increasing for an indefinite distance into space.

Exosphere

The exosphere is the very outer limit of the atmosphere. The bottom of this layer is found at 500 kilometres above the Earth's surface. The pressure drops to little more than a vacuum. Auroras form in the exosphere.

Weather

Weather is the state of the atmosphere at a particular place during a short period of time. It involves such atmospheric phenomena as temperature, humidity, precipitation (type and amount), air pressure, wind, and cloud cover. Weather differs from climate in that the latter includes the synthesis of weather conditions that have prevailed over a given area during a long time period—generally 30 years.

Weather, as most commonly defined, occurs in the troposphere, the lowest region of the atmosphere that extends from the Earth's surface to 6–8 km (4–5 miles) at the poles and to about 17 km (11 miles) at the Equator. Weather is largely confined to the troposphere since this is where almost all clouds occur and almost all precipitation develops. Phenomena occurring in higher regions of the troposphere and above, such as jet streams and upper-air waves, significantly affect sea-level atmospheric-pressure patterns—the so-called highs and lows—and thereby the weather conditions at the terrestrial surface. Geographic features, most notably mountains and large bodies of water (e.g., lakes and oceans), also affect weather. Recent research, for example, has revealed that ocean-surface temperature anomalies are a potential cause of atmospheric temperature anomalies in successive seasons and at distant locations. One manifestation of such weather-affecting interactions between the ocean and the atmosphere is what scientists call the El Niño/Southern Oscillation (ENSO). It is believed that ENSO is responsible not only for unusual weather events in the equatorial Pacific region (e.g., the exceedingly severe drought in Australia and the torrential rains in western South America in 1982–83) but also for those that periodically occur in the mid-latitudes (as, for example, the record-high summer temperatures in western Europe and unusually heavy spring rains in the central United States in 1982–83). The ENSO event of 1997–98 was associated with winter temperatures well above average in much of the United States. The ENSO phenomenon appears to influence mid-latitude weather conditions by modulating the position and intensity of the polar-front jet stream.

Generally speaking, the changeability of weather varies widely in different parts of the world. It is most pronounced in the mid-latitude belts of the westerly winds, where a usually continuous procession of traveling high-and low-pressure centres produces a constantly shifting weather pattern. In tropical regions, by contrast, weather varies little from day to day or from month to month.

Weather has a tremendous influence on human settlement patterns, food production, and personal comfort. Extremes of temperature and humidity cause discomfort and may lead to the transmission of disease; heavy rain can cause flooding, displacing people and interrupting economic activities; thunderstorms, tornadoes, hail, and sleet storms may damage or destroy crops, buildings, and transportation routes and vehicles. Storms may even kill or injure people and livestock. At sea and along adjacent coastal areas, tropical cyclones (also called hurricanes or typhoons) can cause great damage through excessive rainfall and flooding, winds, and wave action to ships, buildings, trees, crops, roads, and railways, and they may interrupt air service and communications. Heavy snowfall and icy conditions can impede transportation and increase the frequency of accidents. The long absence of rainfall, by contrast, can cause droughts and severe dust storms when winds blow over parched farmland, as with the "dustbowl" conditions of the U.S. Plains states in the 1930s.

The variability of weather phenomena has resulted in a long-standing human concern with predictions of future weather conditions and weather forecasting. In early historical times, severe weather was ascribed to annoyed or malevolent divinities. Since the mid-19th century, scientific weather forecasting has evolved, using the precise measurement of air pressure, temperature, humidity, and wind direction and speed to predict changing weather. The development of weather satellites since the 1980s has enabled meteorologists to track the movement of cyclones, anticyclones, their associated fronts, and storms worldwide. In addition, the use of radar permits the monitoring of precipitation, clouds, and tropospheric winds. To predict the weather one week or more in advance, computers combine weather models, which are based on the principles of physics, with measured weather variables, such as current temperature and wind speed. These developments have improved the accuracy of local forecasts and have led to extended and long-range forecasts, although the high variability of weather in the mid-latitudes makes longer-range forecasts less accurate. In tropical regions, by contrast, daily weather variations are small, with regularly occurring phenomena and perceptible change associated more with seasonal cycles (dry weather and monsoons). For some tropical areas, tropical cyclones themselves are one of the more influential weather variables.

Climate

The climate of an area is defined as the aggregate of weather conditions over time, is constructed from monthly, seasonal, and annual averages of weather elements, such as temperature and precipitation, combined with statements about the frequency of extreme events, such as droughts or tornadoes. Historically, climate has had important economic implications for agriculture, transportation, and settlement. Climatology, or the scientific study of climate, dates to the mid-nineteenth century and includes such specialties as applied climatology, climate dynamics, and climate change.

The classical heritage related the climate of an area uniquely to its latitude. Climate, was originally thought to depend only on the height of the sun above the horizon, modified in part by special local characteristics. Climate and health have also been closely related throughout history. According to the Hippocratic tradition of ancient Greece, a physician should consider the seasons of the year and what effects each of them produces; the location of a city with respect to winds, waters, terrain, and the rising of the sun; and the particulars of the weather. These were keys to diagnosing and treating diseases in a given location.

Importance of Meteorology

Meteorology is important because of the impact of air conditions on life. First of all weather forecasting has vital role in urban administration. Cities preparing extreme weather conditions such as tornadoes, snowstorms in order to prevent disasters.

Second, long term weather forecasting is important for agriculture. Drought can cause starvation. Meteorology is important for farmers since crops need water to grow up.

Meteorology is also important for both air and marine transportation. Windy weather effect planes and ships. Lightnings can be dangerous for planes. Sea waves give hard times to ships. Airports stop their operations and ships stay on harbors in extreme weather conditions.

References

- Meteorology, teaching-content, teachers: scholastic.com, Retrieved 20 January, 2019
- Middleschool-atmos, outreach, solveii: solveii, Retrieved 21 February, 2019
- Weather, science: britannica.com, Retrieved 22 March, 2019
- Climate, weather-and-climate-terms-and-concepts, atmosphere-and-weather, earth-and-environment: encyclopedia.com, Retrieved 23 April, 2019
- Why-meteorology-is-important, questions: socratic.org, Retrieved 24 May, 2019

Meteorological Occurrences

2 CHAPTER

The most common weather occurrences which are studied under the domain of meteorology are clouds, rain, water vapor, winds, storms, snow, tropical cyclones, etc. The topics elaborated in this chapter will help in gaining a better perspective about these meteorological occurrences.

CLOUD

Stratocumulus perlucidus clouds, as seen from an airplane window

A cloud is a visible mass of condensed droplets, frozen crystals suspended in the atmosphere above the surface of the Earth or other planetary body, such as a moon. (The term cloud is also applied to a mass of material in interstellar space, where it is called an interstellar cloud or nebula.) The scientific study of clouds is called nephology, which is a branch of meteorology.

On Earth, clouds play an important part in the water cycle (hydrologic cycle). As evaporation occurs from the surface of oceans, lakes, rivers, and other exposed bodies of water, the vapor becomes part of the atmosphere and later condenses to form small droplets or ice crystals, typically 0.01 millimeter (mm) in diameter. When surrounded by billions of other droplets or crystals, they become visible as clouds. Eventually, the water precipitates in the form of rain, snow, and so forth, nourishing the Earth and replenishing the stores of freshwater on land. In addition, clouds regulate the Earth's climate, based on the amount of radiation they reflect and absorb from the Sun and the Earth.

Dense deep clouds exhibit a high reflectance (70-95 percent) throughout the visible range of wavelengths: they thus appear white, at least from the top. Cloud droplets tend to scatter light efficiently, so that the intensity of the solar radiation decreases with depth into the cloud, hence the gray or even sometimes dark appearance of the clouds at their base. Thin clouds may appear to have acquired the color of their environment or background, and clouds illuminated by non-white light, such as during sunrise or sunset, may be colored accordingly. In the near-infrared range, clouds would appear darker because the water that constitutes the cloud droplets strongly absorbs solar radiation at those wavelengths.

Cloud Formation and Properties

Clouds can cast shadows

Clouds may be formed by several different mechanisms.

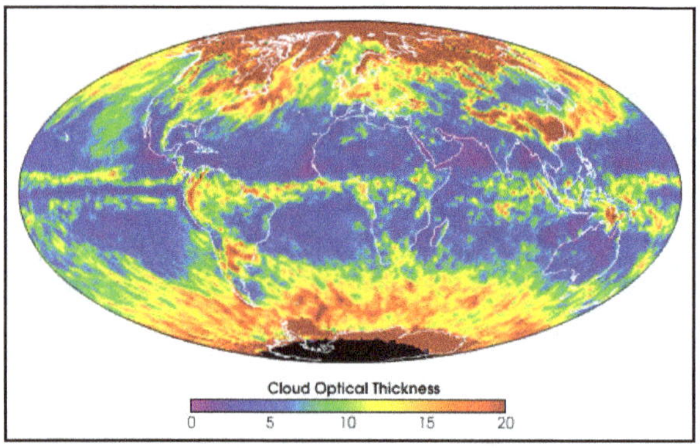

Global scheme of cloud optical thickness

- Water vapor in the air condenses when the air is cooled below its saturation point. This happens when the air comes into contact with a cold surface or a surface that is cooling by radiation, or the air is cooled by adiabatic expansion (rising). This can happen:
    - Along warm and cold fronts (frontal lift);
    - Where air flows up the side of a mountain and cools as it rises (orographic lift);
    - By the convection caused by the warming of a surface by insolation (diurnal heating);
    - When warm air blows over a colder surface, such as a cool body of water.
- Clouds can be formed when two air masses below saturation point mix. Examples are: the condensation of vapor in human breath on a cold day, aircraft contrails, and Arctic sea smoke.

- The air stays the same temperature but absorbs more water vapor into it until it reaches saturation point.

The water in a typical cloud can have a mass of up to several million tonnes. The volume of a cloud is correspondingly high and the net density of the relatively warm air holding the droplets is low enough that air currents below and within the cloud are capable of keeping it suspended.

Conditions inside a cloud are not static: water droplets are constantly forming and re-evaporating. A typical cloud droplet has a radius on the order of 1×10^{-5} m and a terminal velocity of about 1-3 cm/s. This gives these droplets plenty of time to re-evaporate as they fall into the warmer air beneath the cloud.

Most water droplets are formed when water vapor condenses around a condensation nucleus, such as a tiny particle of smoke, dust, ash or salt. In supersaturated conditions, water droplets may act as condensation nuclei.

Water droplets large enough to fall to the ground are produced in two ways. The most important means is through the Bergeron Process, theorized by Tor Bergeron, in which supercooled water droplets and ice crystals in a cloud interact to produce the rapid growth of ice crystals; these crystals precipitate from the cloud and melt as they fall. This process typically takes place in clouds with tops cooler than -15 °C.

The second most important process is the collision and wake capture process, occurring in clouds with warmer tops, in which the collision of rising and falling water droplets produces larger and larger droplets, which are eventually heavy enough to overcome air currents in the cloud and the updraft beneath it and fall as rain. As a droplet falls through the smaller droplets that surround it, it produces a "wake" which draws some of the smaller droplets into collisions, perpetuating the process. This method of raindrop production is the primary mechanism in low stratiform clouds and small cumulus clouds in trade winds and tropical regions and produces raindrops of several millimeters diameter.

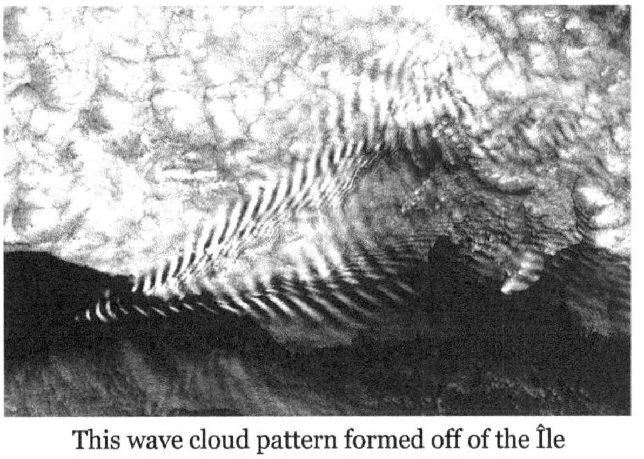

This wave cloud pattern formed off of the Île Amsterdam in the far southern Indian Ocean

The actual form of the cloud created depends on the strength of the uplift and on air stability. In unstable conditions, convection dominates, creating vertically developed clouds. Stable air produces horizontally homogeneous clouds. Frontal uplift creates various cloud forms depending on the composition of the front (ana-type or kata-type warm or cold front). Orographic uplift also creates variable cloud forms depending on air stability, although cap cloud and wave clouds are specific to orographic clouds.

"Hot Ice" and "Ice Memory" in Cloud Formation

In addition to being the colloquial term sometimes used to describe dry ice, "hot ice" is the name given to a surprising phenomenon in which water can be turned into ice at room temperature by supplying an electric field on the order of one million volts per meter.). The effect of such electric fields has been suggested as an explanation of cloud formation. This theory is highly controversial and is not widely accepted as the mechanism of cloud formation. The first time cloud ice forms around a clay particle, it requires a temperature of -10 °C, but subsequent freezing around the same clay particle requires a temperature of just -5 °C, suggesting some kind of "ice memory."

Cloud Classification

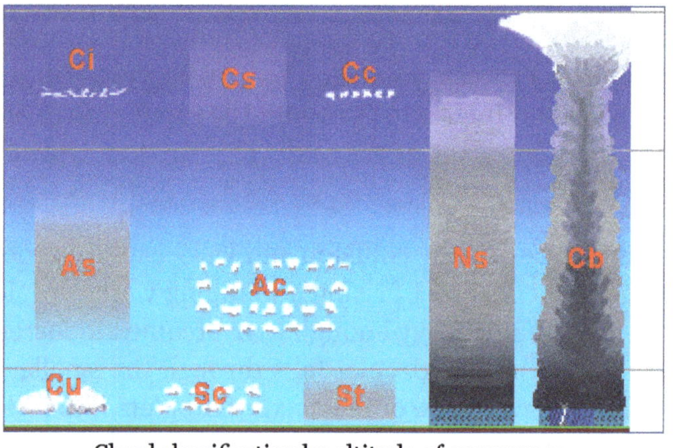

Cloud classification by altitude of occurrence

Clouds are divided into two general categories: layered and convective. These are named stratus clouds and cumulus clouds (or cumuliform; cumulus means "piled up"). These two cloud types are divided into four more groups that distinguish the cloud's altitude. Clouds are classified by the cloud base height, not the cloud top. This system was proposed by Luke Howard in 1802 in a presentation to the Askesian Society.

High Clouds (Family A)

These generally form above 16,500 feet (5,000 m), in the cold region of the troposphere. In Polar regions, they may form as low as 10,000 ft (3,048 m); they are denoted by the prefix cirro- or cirrus. At this altitude, water frequently freezes so clouds are composed of ice crystals. The clouds tend to be wispy and are often transparent.

Clouds in Family A include:

- Cirrus (CI).
- Cirrus uncinus.
- Cirrus Kelvin-Helmholtz Colombia.

- Cirrostratus (Cs).
- Cirrocumulus (Cc).
- Pileus.
- Contrail, a long thin cloud which develops as the result of the passage of an aircraft at high altitudes.

Middle Clouds (Family B)

Altocumulus mackerel sky

These develop between 6,500 and 16,500 feet (between 2,000 and 5,000 m) and are denoted by the prefix alto-. They are made of water droplets and are frequently supercooled.

Clouds in Family B include:

- Altostratus (As).
- Altostratus undulatus.
- Altocumulus (Ac).
- Altocumulus undulatus.
- Altocumulus mackerel sky.
- Altocumulus castellanus.
- Altocumulus lenticularis.

Low Clouds (Family C)

These are found up to 6,500 feet (2,000 m) and include the stratus (dense and grey). When stratus clouds contact the ground, they are called fog.

Clouds in Family C include:

- Stratus (St).

- Nimbostratus (Ns).
- Cumulus humilis (Cu).
- Cumulus mediocris (Cu).
- Stratocumulus (Sc).

Low clouds

Vertical Clouds (Family D)

Cumulonimbus clouds showing strong updrafts

These clouds can have strong up-currents, rise far above their bases and form at many heights.

Clouds in Family D include:

- Cumulonimbus (associated with heavy precipitation and thunderstorms) (Cb).
- Cumulonimbus incus.
- Cumulonimbus calvus.
- Cumulonimbus with mammatus.
- Cumulus congestus.
- Pyrocumulus.

Mammatus cloud formations

Other Clouds

A few clouds can be found above the troposphere; these include noctilucent and polar stratospheric clouds (or nacreous clouds), which occur in the mesosphere and stratosphere respectively.

Cloud Fields

A cloud field is simply a group of clouds but sometimes cloud fields can take on certain shapes that have their own characteristics and are specially classified. Stratocumulus clouds can often be found in the following forms:

- Open cell, which resembles a honeycomb, with clouds around the edges and clear, open space in the middle.
- Closed cell, which is cloudy in the center and clear on the edges, similar to a filled honeycomb.
- Actinoform, which resembles a leaf or a spoked wheel.

Colors

Iridescent clouds

An example of various cloud colors

Rain bearing clouds

The color of a cloud tells much about what is going on inside the cloud. Clouds form when relatively warm air containing water vapor is lighter than its surrounding air and this causes it to rise. As it rises it cools and the vapor condenses out of the air as micro-droplets. These tiny particles of water are relatively densely packed and sunlight cannot penetrate far into the cloud before it is reflected out, giving a cloud its characteristic white color. As a cloud matures, the droplets may combine to produce larger droplets, which may combine to form droplets large enough to fall as rain. In this process of accumulation, the space between droplets becomes larger and larger, permitting light to penetrate much farther into the cloud. If the cloud is sufficiently large and the droplets within are spaced far enough apart, it may be that a percentage of the light which enters the cloud is not reflected back out before it is absorbed (Think of how much farther one can see in a heavy rain as opposed to how far one can see in a heavy fog). This process of reflection/absorption is what leads to the range of cloud color from white through gray through black. For the same reason, the undersides of large clouds and heavy overcasts appear various degrees of gray; little light is being reflected or transmitted back to the observer.

Other colors occur naturally in clouds. Bluish-gray is the result of light scattering within the cloud. In the visible spectrum, blue and green are at the short end of light's visible wavelengths, while red and yellow are at the long end. The short rays are more easily scattered by water droplets, and the long rays are more likely to be absorbed. The bluish color is evidence that such scattering is being produced by rain-sized droplets in the cloud.

A greenish tinge to a cloud is produced when sunlight is scattered by ice. A cumulonimbus cloud which shows green is a pretty sure sign of imminent heavy rain, hail, strong winds and possible tornadoes.

Yellowish clouds are rare but may occur in the late spring through early fall months during forest fire season. The yellow color is due to the presence of smoke.

Red, orange, and pink clouds occur almost entirely at sunrise/sunset and are the result of the scattering of sunlight by the atmosphere. The clouds are not that color; they are reflecting the long (and unscattered) rays of sunlight which are predominant at those hours. The effect is much the same as if one were to shine a red spotlight on a white sheet. In combination with large, mature thunderheads this can produce blood-red clouds. The evening before the Edmonton, Alberta tornado in 1987, Edmontonians observed such clouds — deep black on their dark side and intense red on their sunward side. In this case the adage "red sky at night, sailor's delight" was wrong.

Global Dimming

The recently recognized phenomenon of global dimming is thought to be caused by changes to the reflectivity of clouds due to the increased presence of aerosols and other particulates in the atmosphere.

Global Brightening

Recent research by Martin Wild and coworkers indicates a global brightening trend. Global brightening is caused by decreased amounts of particulate matter in the atmosphere. With less particulate matter, there is less surface area for condensation to occur. Since there's less condensation in the atmosphere and increased evaporation caused by increasing amounts of sunlight striking the water's surface there is more moisture, causing fewer but thicker clouds.

Clouds on other Planets

Within our Solar System, any planet or moon with an atmosphere also has clouds. Venus' clouds are composed entirely of sulfuric acid droplets. Mars has high, thin clouds of water ice. Both Jupiter and Saturn have an outer cloud deck composed of ammonia clouds, an intermediate deck of ammonium hydrosulfide clouds and an inner deck of water clouds. Uranus and Neptune have atmospheres dominated by methane clouds.

Saturn's moon Titan has clouds which are believed to be composed largely of droplets of liquid methane. The Cassini-Huygens Saturn mission has uncovered evidence of a fluid cycle on Titan, including lakes near the poles and fluvial channels on the surface of the moon.

RAIN

Rain is liquid water in the form of droplets that have condensed from atmospheric water vapor and then become heavy enough to fall under gravity. Rain is a major component of the water cycle and is responsible for depositing most of the fresh water on the Earth. It provides suitable conditions for many types of ecosystems, as well as water for hydroelectric power plants and crop irrigation.

The major cause of rain production is moisture moving along three-dimensional zones of temperature and moisture contrasts known as weather fronts. If enough moisture and upward motion is present, precipitation falls from convective clouds (those with strong upward vertical motion) such as cumulonimbus (thunder clouds) which can organize into narrow rainbands. In mountainous areas, heavy precipitation is possible where upslope flow is maximized within windward sides of the terrain at elevation which forces moist air to condense and fall out as rainfall along the sides of mountains. On the leeward side of mountains, desert climates can exist due to the dry air caused by downslope flow which causes heating and drying of the air mass. The movement of the monsoon trough, or intertropical convergence zone, brings rainy seasons to savannah climes.

Heavy rain falling over the desert at sunset

The urban heat island effect leads to increased rainfall, both in amounts and intensity, downwind of cities. Global warming is also causing changes in the precipitation pattern globally, including wetter conditions across eastern North America and drier conditions in the tropics. Antarctica is the driest continent. The globally averaged annual precipitation over land is 715 mm (28.1 in), but over the whole Earth it is much higher at 990 mm (39 in). Climate classification systems such as the Köppen classification system use average annual rainfall to help differentiate between differing climate regimes. Rainfall is measured using rain gauges. Rainfall amounts can be estimated by weather radar.

Rain is also known or suspected on other planets, where it may be composed of methane, neon, sulfuric acid, or even iron rather than water.

Formation

Water-saturated Air

Rain falling on a field, in southern Estonia

Air contains water vapor, and the amount of water in a given mass of dry air, known as the *mixing ratio*, is measured in grams of water per kilogram of dry air (g/kg). The amount of moisture in air is also commonly reported as relative humidity; which is the percentage of the total water vapor air can hold at a particular air temperature. How much water vapor a parcel of air can contain before it becomes saturated (100% relative humidity) and forms into a cloud (a group of visible and tiny water and ice particles suspended above the Earth's surface) depends on its temperature. Warmer air can contain more water vapor than cooler air before becoming saturated. Therefore, one way to saturate a parcel of air is to cool it. The dew point is the temperature to which a parcel must be cooled in order to become saturated.

There are four main mechanisms for cooling the air to its dew point: adiabatic cooling, conductive cooling, radiational cooling, and evaporative cooling. Adiabatic cooling occurs when air rises and expands. The air can rise due to convection, large-scale atmospheric motions, or a physical barrier such as a mountain (orographic lift). Conductive cooling occurs when the air comes into contact with a colder surface, usually by being blown from one surface to another, for example from a liquid water surface to colder land. Radiational cooling occurs due to the emission of infrared radiation, either by the air or by the surface underneath. Evaporative cooling occurs when moisture is added to the air through evaporation, which forces the air temperature to cool to its wet-bulb temperature, or until it reaches saturation.

The main ways water vapor is added to the air are: wind convergence into areas of upward motion, precipitation or virga falling from above, daytime heating evaporating water from the surface of oceans, water bodies or wet land, transpiration from plants, cool or dry air moving over warmer water, and lifting air over mountains. Water vapor normally begins to condense on condensation nuclei such as dust, ice, and salt in order to form clouds. Elevated portions of weather fronts (which are three-dimensional in nature) force broad areas of upward motion within the Earth's atmosphere which form clouds decks such as altostratus or cirrostratus. Stratus is a stable cloud deck which tends to form when a cool, stable air mass is trapped underneath a warm air mass. It can also form due to the lifting of advection fog during breezy conditions.

Coalescence and Fragmentation

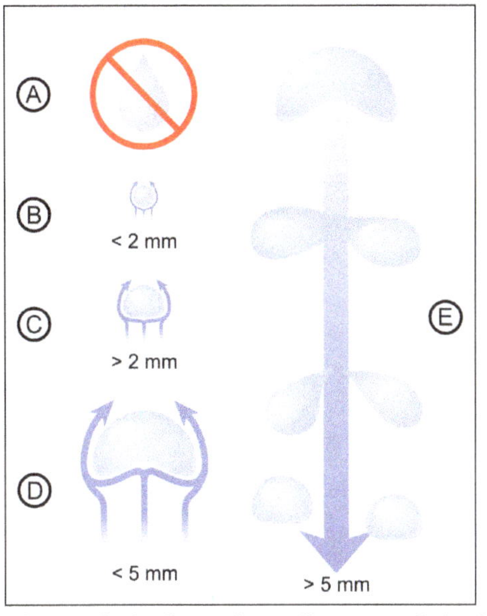

The shape of rain drops depending upon their size

Coalescence occurs when water droplets fuse to create larger water droplets. Air resistance typically causes the water droplets in a cloud to remain stationary. When air turbulence occurs, water droplets collide, producing larger droplets.

Black Rain Clouds

As these larger water droplets descend, coalescence continues, so that drops become heavy enough to overcome air resistance and fall as rain. Coalescence generally happens most often in clouds above freezing, and is also known as the warm rain process. In clouds below freezing, when ice crystals gain enough mass they begin to fall. This generally requires more mass than coalescence when occurring between the crystal and neighboring water droplets. This process is temperature dependent, as supercooled water droplets only exist in a cloud that is below freezing. In addition, because of the great temperature difference between cloud and ground level, these ice crystals may melt as they fall and become rain.

Raindrops have sizes ranging from 0.1 to 9 mm (0.0039 to 0.3543 in) mean diameter but develop a tendency to break up at larger sizes. Smaller drops are called cloud droplets, and their shape is spherical. As a raindrop increases in size, its shape becomes more oblate, with its largest cross-section facing the oncoming airflow. Large rain drops become increasingly flattened on the bottom, like hamburger buns; very large ones are shaped like parachutes. Contrary to popular belief, their shape does not resemble a teardrop. The biggest raindrops on Earth were recorded over Brazil and the Marshall Islands in 2004 – some of them were as large as 10 mm (0.39 in). The large size is explained by condensation on large smoke particles or by collisions between drops in small regions with particularly high content of liquid water.

Rain drops associated with melting hail tend to be larger than other rain drops.

A raindrop on a leaf

Intensity and duration of rainfall are usually inversely related, i.e., high intensity storms are likely to be of short duration and low intensity storms can have a long duration.

Droplet Size Distribution

The final droplet size distribution is an exponential distribution. The number of droplets with diameter between d and $D + dD$ per unit volume of space is $n(d) = n_0 e^{-d/\langle d \rangle} dD$. This is commonly referred to as the Marshall–Palmer law after the researchers who first characterized it. The parameters are somewhat temperature-dependent, and the slope also scales with the rate of rainfall $\langle d \rangle^{-1} = 41R^{-0.21}$ (d in centimeters and R in millimetres per hour).

Deviations can occur for small droplets and during different rainfall conditions. The distribution tends to fit averaged rainfall, while instantaneous size spectra often deviate and have been modeled as gamma distributions. The distribution has an upper limit due to droplet fragmentation.

Raindrop Impacts

Raindrops impact at their terminal velocity, which is greater for larger drops due to their larger mass to drag ratio. At sea level and without wind, 0.5 mm (0.020 in) drizzle impacts at 2 m/s (6.6 ft/s) or 7.2 km/h (4.5 mph), while large 5 mm (0.20 in) drops impact at around 9 m/s (30 ft/s) or 32 km/h (20 mph).

Rain falling on loosely packed material such as newly fallen ash can produce dimples that can be fossilized, called raindrop impressions. The air density dependence of the maximum raindrop diameter together with fossil raindrop imprints has been used to constrain the density of the air 2.7 billion years ago.

The sound of raindrops hitting water is caused by bubbles of air oscillating underwater. The METAR code for rain is RA, while the coding for rain showers is SHRA.

Virga

In certain conditions precipitation may fall from a cloud but then evaporate or sublime before reaching the ground. This is termed virga and is more often seen in hot and dry climates.

Causes

Frontal Activity

Stratiform (a broad shield of precipitation with a relatively similar intensity) and dynamic precipitation (convective precipitation which is showery in nature with large changes in intensity over short distances) occur as a consequence of slow ascent of air in synoptic systems (on the order of cm/s), such as in the vicinity of cold fronts and near and poleward of surface warm fronts. Similar ascent is seen around tropical cyclones outside the eyewall, and in comma-head precipitation patterns around mid-latitude cyclones. A wide variety of weather can be found along an occluded front, with thunderstorms possible, but usually their passage is associated with a drying of the air mass. Occluded fronts usually form around mature low-pressure areas. What separates rainfall from other precipitation types, such as ice pellets and snow, is the presence of a thick layer of air aloft which is above the melting point of water, which melts the frozen precipitation well before it reaches the ground. If there is a shallow near surface layer that is below freezing, freezing rain (rain which freezes on contact with surfaces in subfreezing environments) will result. Hail becomes an increasingly infrequent occurrence when the freezing level within the atmosphere exceeds 3,400 m (11,000 ft) above ground level.

Convection

Convective rain, or showery precipitation, occurs from convective clouds (e.g., cumulonimbus or cumulus congestus). It falls as showers with rapidly changing intensity. Convective precipitation falls over a certain area for a relatively short time, as convective clouds have limited horizontal extent. Most precipitation in the tropics appears to be convective; however, it has been suggested that stratiform precipitation also occurs. Graupel and hail indicate convection. In mid-latitudes, convective precipitation is intermittent and often associated with baroclinic boundaries such as cold fronts, squall lines, and warm fronts.

Convective precipitation

Orographic precipitation

Orographic Effects

Orographic precipitation occurs on the windward side of mountains and is caused by the rising air motion of a large-scale flow of moist air across the mountain ridge, resulting in adiabatic cooling and condensation. In mountainous parts of the world subjected to relatively consistent winds (for example, the trade winds), a more moist climate usually prevails on the windward side of a mountain than on the leeward or downwind side. Moisture is removed by orographic lift, leaving drier air on the descending and generally warming, leeward side where a rain shadow is observed.

In Hawaii, Mount Wai'ale'ale, on the island of Kauai, is notable for its extreme rainfall, as it is amongst the places in the world with the highest levels of rainfall, with 9,500 mm (373 in). Systems known as Kona storms affect the state with heavy rains between October and April. Local climates vary considerably on each island due to their topography, divisible into windward (*Ko'olau*) and leeward (*Kona*) regions based upon location relative to the higher mountains. Windward sides face the east to northeast trade winds and receive much more rainfall; leeward sides are drier and sunnier, with less rain and less cloud cover.

In South America, the Andes mountain range blocks Pacific moisture that arrives in that continent, resulting in a desertlike climate just downwind across western Argentina. The Sierra Nevada range creates the same effect in North America forming the Great Basin and Mojave Deserts.

Within the Tropics

The wet, or rainy, season is the time of year, covering one or more months, when most of the average annual rainfall in a region falls. The term *green season* is also sometimes used as a euphemism by tourist authorities. Areas with wet seasons are dispersed across portions of the tropics and subtropics. Savanna climates and areas with monsoon regimes have wet summers and dry winters. Tropical rainforests technically do not have dry or wet seasons, since their rainfall is equally distributed through the year. Some areas with pronounced rainy seasons will see a break in rainfall mid-season when the intertropical convergence zone or monsoon trough move poleward of their location during the middle of the warm season. When the wet season occurs during the warm season, or summer, rain falls mainly during the late afternoon and early evening hours. The wet season is a time when air quality improves, freshwater quality improves, and vegetation grows significantly.

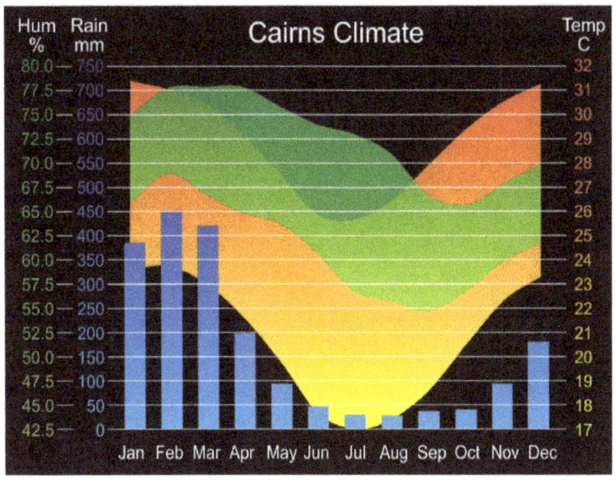

Rainfall distribution by month in Cairns showing the extent of the wet season at that location

Tropical cyclones, a source of very heavy rainfall, consist of large air masses several hundred miles across with low pressure at the centre and with winds blowing inward towards the centre in either a clockwise direction (southern hemisphere) or counter clockwise (northern hemisphere). Although cyclones can take an enormous toll in lives and personal property, they may be important factors in the precipitation regimes of places they impact, as they may bring much-needed precipitation to otherwise dry regions. Areas in their path can receive a year's worth of rainfall from a tropical cyclone passage.

Human Influence

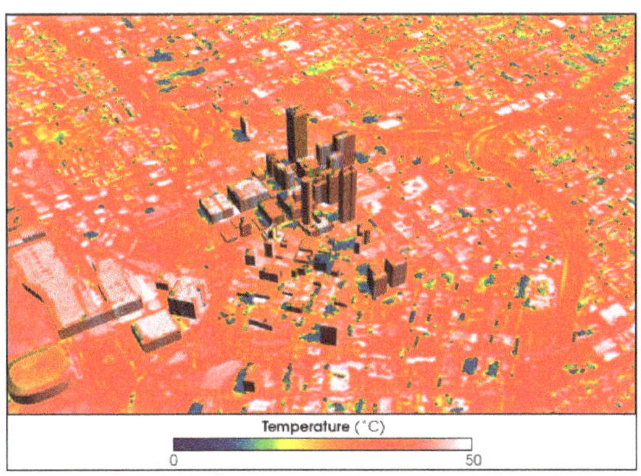

Image of Atlanta, Georgia showing temperature distribution, with blue showing cool temperatures, red warm, and hot areas appearing white

The fine particulate matter produced by car exhaust and other human sources of pollution forms cloud condensation nuclei, leads to the production of clouds and increases the likelihood of rain. As commuters and commercial traffic cause pollution to build up over the course of the week, the likelihood of rain increases: it peaks by Saturday, after five days of weekday pollution has been built up. In heavily populated areas that are near the coast, such as the United States' Eastern Seaboard, the effect can be dramatic: there is a 22% higher chance of rain on Saturdays than on Mondays. The urban heat island effect warms cities 0.6 to 5.6 °C (1.1 to 10.1 °F) above surrounding suburbs and rural areas. This extra heat leads to greater upward motion, which can induce additional shower and thunderstorm activity. Rainfall rates downwind of cities are increased between

48% and 116%. Partly as a result of this warming, monthly rainfall is about 28% greater between 32 to 64 km (20 to 40 mi) downwind of cities, compared with upwind. Some cities induce a total precipitation increase of 51%.

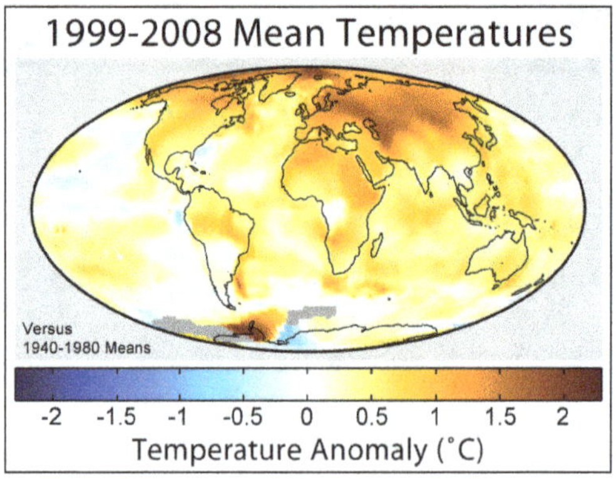

Mean surface temperature anomalies during the period 1999 to 2008 with respect to the average temperatures from 1940 to 1980

Increasing temperatures tend to increase evaporation which can lead to more precipitation. Precipitation generally increased over land north of 30°N from 1900 through 2005 but has declined over the tropics since the 1970s. Globally there has been no statistically significant overall trend in precipitation over the past century, although trends have varied widely by region and over time. Eastern portions of North and South America, northern Europe, and northern and central Asia have become wetter. The Sahel, the Mediterranean, southern Africa and parts of southern Asia have become drier. There has been an increase in the number of heavy precipitation events over many areas during the past century, as well as an increase since the 1970s in the prevalence of droughts—especially in the tropics and subtropics. Changes in precipitation and evaporation over the oceans are suggested by the decreased salinity of mid- and high-latitude waters (implying more precipitation), along with increased salinity in lower latitudes (implying less precipitation and/or more evaporation). Over the contiguous United States, total annual precipitation increased at an average rate of 6.1 percent since 1900, with the greatest increases within the East North Central climate region (11.6 percent per century) and the South (11.1 percent). Hawaii was the only region to show a decrease (−9.25 percent).

Analysis of 65 years of United States of America rainfall records show the lower 48 states have an increase in heavy downpours since 1950. The largest increases are in the Northeast and Midwest, which in the past decade, have seen 31 and 16 percent more heavy downpours compared to the 1950s. Rhode Island is the state with the largest increase, 104%. McAllen, Texas is the city with the largest increase, 700%. Heavy downpour in the analysis are the days where total precipitation exceeded the top 1 percent of all rain and snow days during the years 1950–2014.

The most successful attempts at influencing weather involve cloud seeding, which include techniques used to increase winter precipitation over mountains and suppress hail.

Characteristics of Rain

Patterns

Rainbands are cloud and precipitation areas which are significantly elongated. Rainbands can be stratiform or convective, and are generated by differences in temperature. When noted on weather radar imagery, this precipitation elongation is referred to as banded structure. Rainbands in advance of warm occluded fronts and warm fronts are associated with weak upward motion, and tend to be wide and stratiform in nature.

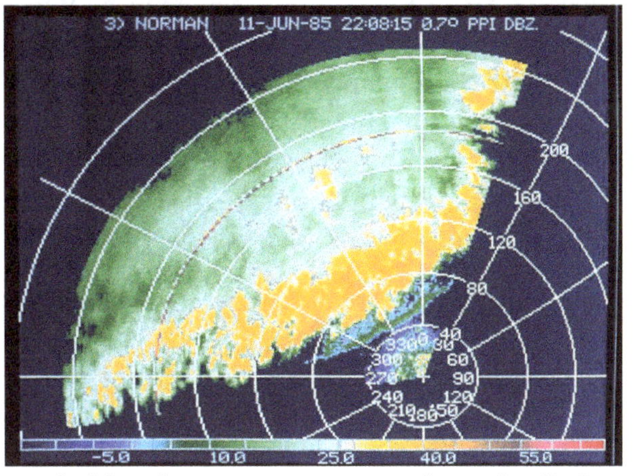

Band of thunderstorms seen on a weather radar display

Rainbands spawned near and ahead of cold fronts can be squall lines which are able to produce tornadoes. Rainbands associated with cold fronts can be warped by mountain barriers perpendicular to the front's orientation due to the formation of a low-level barrier jet. Bands of thunderstorms can form with sea breeze and land breeze boundaries, if enough moisture is present. If sea breeze rainbands become active enough just ahead of a cold front, they can mask the location of the cold front itself.

Once a cyclone occludes, a trough of warm air aloft, or "trowal" for short, will be caused by strong southerly winds on its eastern periphery rotating aloft around its northeast, and ultimately northwestern, periphery (also known as the warm conveyor belt), forcing a surface trough to continue into the cold sector on a similar curve to the occluded front. The trowal creates the portion of an occluded cyclone known as its comma head, due to the comma-like shape of the mid-tropospheric cloudiness that accompanies the feature. It can also be the focus of locally heavy precipitation, with thunderstorms possible if the atmosphere along the trowal is unstable enough for convection. Banding within the comma head precipitation pattern of an extratropical cyclone can yield significant amounts of rain. Behind extratropical cyclones during fall and winter, rainbands can form downwind of relative warm bodies of water such as the Great Lakes. Downwind of islands, bands of showers and thunderstorms can develop due to low level wind convergence downwind of the island edges. Offshore California, this has been noted in the wake of cold fronts.

Rainbands within tropical cyclones are curved in orientation. Tropical cyclone rainbands contain showers and thunderstorms that, together with the eyewall and the eye, constitute a hurricane or tropical storm. The extent of rainbands around a tropical cyclone can help determine the cyclone's intensity.

Acidity

The phrase *acid rain* was first used by Scottish chemist Robert Augus Smith in 1852. The pH of rain varies, especially due to its origin. On America's East Coast, rain that is derived from the

Atlantic Ocean typically has a pH of 5.0–5.6; rain that comes across the continental from the west has a pH of 3.8–4.8; and local thunderstorms can have a pH as low as 2.0. Rain becomes acidic primarily due to the presence of two strong acids, sulfuric acid (H_2SO_4) and nitric acid (HNO_3). Sulfuric acid is derived from natural sources such as volcanoes, and wetlands (sulfate reducing bacteria); and anthropogenic sources such as the combustion of fossil fuels, and mining where H_2S is present. Nitric acid is produced by natural sources such as lightning, soil bacteria, and natural fires; while also produced anthropogenically by the combustion of fossil fuels and from power plants. In the past 20 years the concentrations of nitric and sulfuric acid has decreased in presence of rainwater, which may be due to the significant increase in ammonium (most likely as ammonia from livestock production), which acts as a buffer in acid rain and raises the pH.

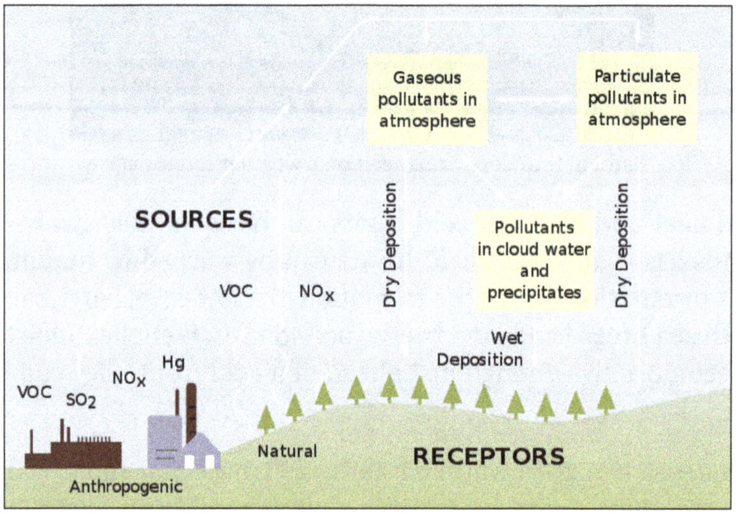

Sources of acid rain

Köppen Climate Classification

The Köppen classification depends on average monthly values of temperature and precipitation. The most commonly used form of the Köppen classification has five primary types labeled A through E. Specifically, the primary types are A, tropical; B, dry; C, mild mid-latitude; D, cold mid-latitude; and E, polar. The five primary classifications can be further divided into secondary classifications such as rain forest, monsoon, tropical savanna, humid subtropical, humid continental, oceanic climate, Mediterranean climate, steppe, subarctic climate, tundra, polar ice cap, and desert.

Rain forests are characterized by high rainfall, with definitions setting minimum normal annual rainfall between 1,750 and 2,000 mm (69 and 79 in). A tropical savanna is a grassland biome located in semi-arid to semi-humid climate regions of subtropical and tropical latitudes, with rainfall between 750 and 1,270 mm (30 and 50 in) a year. They are widespread on Africa, and are also found in India, the northern parts of South America, Malaysia, and Australia. The humid subtropical climate zone is where winter rainfall is associated with large storms that the westerlies steer from west to east. Most summer rainfall occurs during thunderstorms and from occasional tropical cyclones. Humid subtropical climates lie on the east side continents, roughly between latitudes 20° and 40° degrees away from the equator.

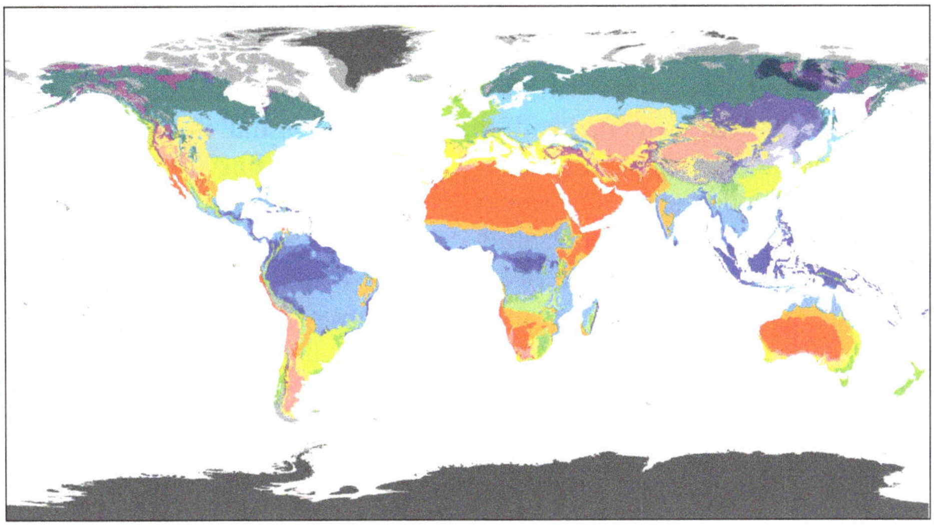

Updated Köppen-Geiger climate map

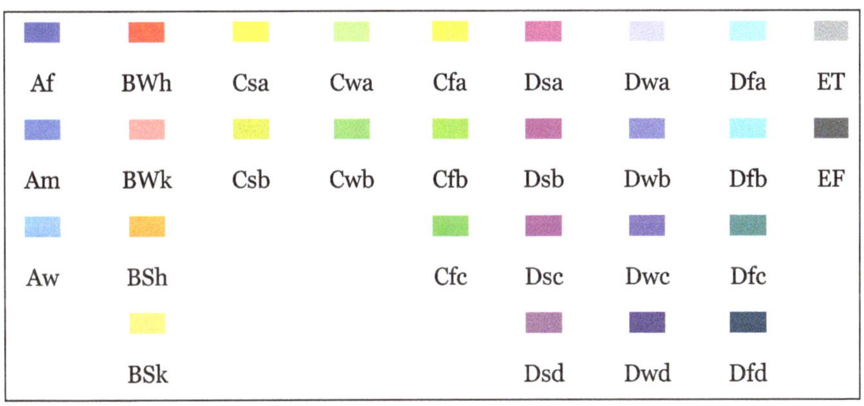

An oceanic (or maritime) climate is typically found along the west coasts at the middle latitudes of all the world's continents, bordering cool oceans, as well as southeastern Australia, and is accompanied by plentiful precipitation year-round. The Mediterranean climate regime resembles the climate of the lands in the Mediterranean Basin, parts of western North America, parts of Western and South Australia, in southwestern South Africa and in parts of central Chile. The climate is characterized by hot, dry summers and cool, wet winters. A steppe is a dry grassland. Subarctic climates are cold with continuous permafrost and little precipitation.

Measurement

Gauges

Rain is measured in units of length per unit time, typically in millimeters per hour, or in countries where imperial units are more common, inches per hour. The "length", or more accurately, "depth" being measured is the depth of rain water that would accumulate on a flat, horizontal and impermeable surface during a given amount of time, typically an hour. One millimeter of rainfall is the equivalent of one liter of water per square meter.

The standard way of measuring rainfall or snowfall is the standard rain gauge, which can be found in 100-mm (4-in) plastic and 200-mm (8-in) metal varieties. The inner cylinder is filled by 25 mm

(0.98 in) of rain, with overflow flowing into the outer cylinder. Plastic gauges have markings on the inner cylinder down to 0.25 mm (0.0098 in) resolution, while metal gauges require use of a stick designed with the appropriate 0.25 mm (0.0098 in) markings. After the inner cylinder is filled, the amount inside it is discarded, then filled with the remaining rainfall in the outer cylinder until all the fluid in the outer cylinder is gone, adding to the overall total until the outer cylinder is empty. Other types of gauges include the popular wedge gauge (the cheapest rain gauge and most fragile), the tipping bucket rain gauge, and the weighing rain gauge. For those looking to measure rainfall the most inexpensively, a can that is cylindrical with straight sides will act as a rain gauge if left out in the open, but its accuracy will depend on what ruler is used to measure the rain with. Any of the above rain gauges can be made at home, with enough know-how.

Standard rain gauge

When a precipitation measurement is made, various networks exist across the United States and elsewhere where rainfall measurements can be submitted through the Internet, such as CoCo-RAHS or GLOBE. If a network is not available in the area where one lives, the nearest local weather or met office will likely be interested in the measurement.

Remote Sensing

One of the main uses of weather radar is to be able to assess the amount of precipitations fallen over large basins for hydrological purposes. For instance, river flood control, sewer management and dam construction are all areas where planners use rainfall accumulation data. Radar-derived rainfall estimates compliment surface station data which can be used for calibration. To produce radar accumulations, rain rates over a point are estimated by using the value of reflectivity data at individual grid points. A radar equation is then used, which is,

$$Z = AR^b,$$

where Z represents the radar reflectivity, R represents the rainfall rate, and A and b are constants. Satellite derived rainfall estimates use passive microwave instruments aboard polar orbiting as well as geostationary weather satellites to indirectly measure rainfall rates. If one wants an accumulated rainfall over a time period, one has to add up all the accumulations from each grid box within the images during that time.

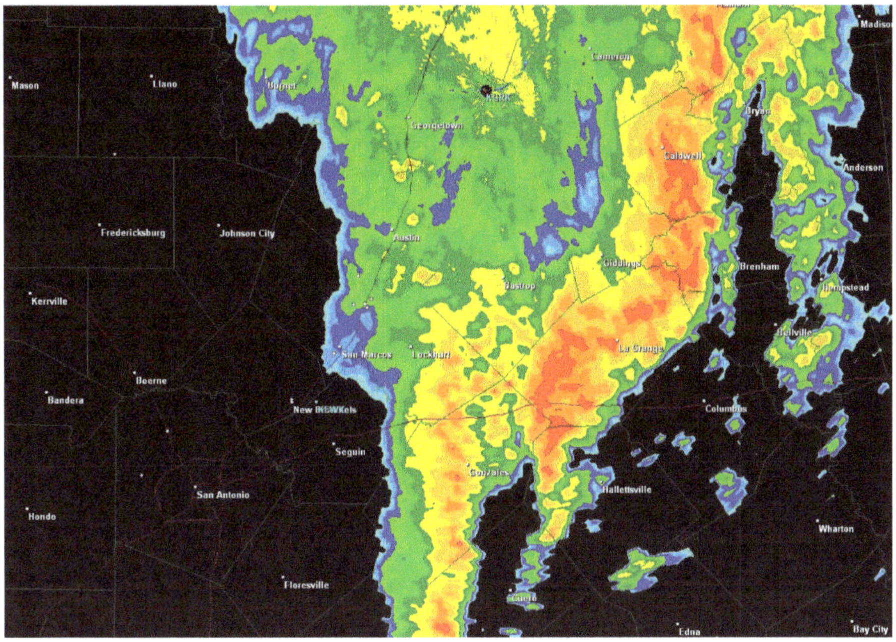

Twenty-four-hour rainfall accumulation on the Val d'Irène radar in Eastern Canada. Zones without data in the east and southwest are caused by beam blocking from mountains

Intensity

Rainfall intensity is classified according to the rate of precipitation, which depends on the considered time. The following categories are used to classify rainfall intensity:

- Light rain — when the precipitation rate is < 2.5 mm (0.098 in) per hour.
- Moderate rain — when the precipitation rate is between 2.5 mm (0.098 in) - 7.6 mm (0.30 in) or 10 mm (0.39 in) per hour.
- Heavy rain — when the precipitation rate is > 7.6 mm (0.30 in) per hour, or between 10 mm (0.39 in) and 50 mm (2.0 in) per hour.
- Violent rain — when the precipitation rate is > 50 mm (2.0 in) per hour.

Euphemisms for a heavy or violent rain include gully washer, trash-mover and toad-strangler. The intensity can also be expressed by rainfall erosivity *R-factor* or in terms of the rainfall time-structure *n-index*.

Return Period

The likelihood or probability of an event with a specified intensity and duration, is called the return period or frequency. The intensity of a storm can be predicted for any return period and storm duration, from charts based on historic data for the location. The term *1 in 10 year storm* describes a rainfall event which is unusual and has a 50% chance of occurring in any 10-year period. The term *1 in 100 year storm* describes a rainfall event which is rare and which will occur with a 50% probability in any 100-year period. As with all probabilistic events, it is possible, though improbable, to have multiple "1 in 100 Year Storms" in a single year.

Forecasting

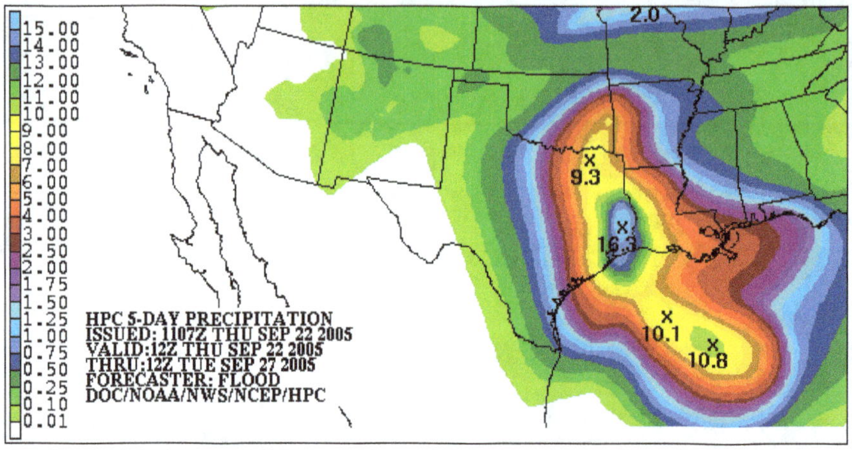

Example of a five-day rainfall forecast from the Hydrometeorological Prediction Center

The Quantitative Precipitation Forecast (abbreviated QPF) is the expected amount of liquid precipitation accumulated over a specified time period over a specified area. A QPF will be specified when a measurable precipitation type reaching a minimum threshold is forecast for any hour during a QPF valid period. Precipitation forecasts tend to be bound by synoptic hours such as 0000, 0600, 1200 and 1800 GMT. Terrain is considered in QPFs by use of topography or based upon climatological precipitation patterns from observations with fine detail. Starting in the mid to late 1990s, QPFs were used within hydrologic forecast models to simulate impact to rivers throughout the United States. Forecast models show significant sensitivity to humidity levels within the planetary boundary layer, or in the lowest levels of the atmosphere, which decreases with height. QPF can be generated on a quantitative, forecasting amounts, or a qualitative, forecasting the probability of a specific amount, basis. Radar imagery forecasting techniques show higher skill than model forecasts within 6 to 7 hours of the time of the radar image. The forecasts can be verified through use of rain gauge measurements, weather radar estimates, or a combination of both. Various skill scores can be determined to measure the value of the rainfall forecast.

Impact

Agricultural

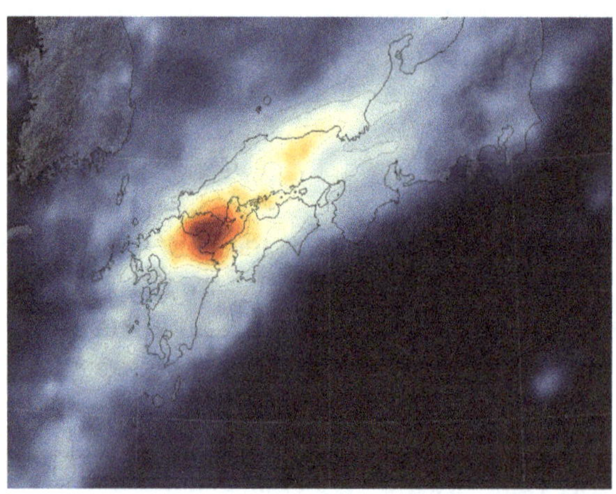

Rainfall for southern Japan and the surrounding region

Precipitation, especially rain, has a dramatic effect on agriculture. All plants need at least some water to survive, therefore rain (being the most effective means of watering) is important to agriculture. While a regular rain pattern is usually vital to healthy plants, too much or too little rainfall can be harmful, even devastating to crops. Drought can kill crops and increase erosion, while overly wet weather can cause harmful fungus growth. Plants need varying amounts of rainfall to survive. For example, certain cacti require small amounts of water, while tropical plants may need up to hundreds of inches of rain per year to survive.

In areas with wet and dry seasons, soil nutrients diminish and erosion increases during the wet season. Animals have adaptation and survival strategies for the wetter regime. The previous dry season leads to food shortages into the wet season, as the crops have yet to mature. Developing countries have noted that their populations show seasonal weight fluctuations due to food shortages seen before the first harvest, which occurs late in the wet season. Rain may be harvested through the use of rainwater tanks; treated to potable use or for non-potable use indoors or for irrigation. Excessive rain during short periods of time can cause flash floods.

Cultural and Religious

Rain dance being performed in Harar, Ethiopia

Cultural attitudes towards rain differ across the world. In temperate climates, people tend to be more stressed when the weather is unstable or cloudy, with its impact greater on men than women. Rain can also bring joy, as some consider it to be soothing or enjoy the aesthetic appeal of it. In dry places, such as India, or during periods of drought, rain lifts people's moods. In Botswana, the Setswana word for rain, *pula*, is used as the name of the national currency, in recognition of the economic importance of rain in its country, since it has a desert climate. Several cultures have developed means of dealing with rain and have developed numerous protection devices such as umbrellas and raincoats, and diversion devices such as gutters and storm drains that lead rains to sewers. Many people find the scent during and immediately after rain pleasant or distinctive. The source of this scent is petrichor, an oil produced by plants, then absorbed by rocks and soil, and later released into the air during rainfall.

Rain holds an important religious significance in many cultures. The ancient Sumerians believed that rain was the semen of the sky-god An, which fell from the heavens to inseminate his consort, the earth-goddess Ki, causing her to give birth to all the plants of the earth. The Akkadians believed that the clouds were the breasts of Anu's consort Antu and that rain was milk from her breasts. According to Jewish tradition, in the first century BC, the Jewish miracle-worker Honi ha-M'agel ended a three-year drought in Judaea by drawing a circle in the sand and praying for rain, refusing to leave the circle until his prayer was granted. In his *Meditations*, the Roman emperor Marcus Aurelius preserves a prayer for rain made by the Athenians to the Greek sky-god Zeus. Various Native American tribes are known to have historically conducted rain dances in effort to encourage rainfall. Rainmaking rituals are also important in many African cultures. In the present-day United States, various state governors have held Days of Prayer for rain, including the Days of Prayer for Rain in the State of Texas in 2011.

Rain, depicted in the 1493 *Nuremberg Chronicle*

Global Climatology

Approximately 505,000 km³ (121,000 cu mi) of water falls as precipitation each year across the globe with 398,000 km³ (95,000 cu mi) of it over the oceans. Given the Earth's surface area, that means the globally averaged annual precipitation is 990 mm (39 in). Deserts are defined as areas with an average annual precipitation of less than 250 mm (10 in) per year, or as areas where more water is lost by evapotranspiration than falls as precipitation.

Deserts

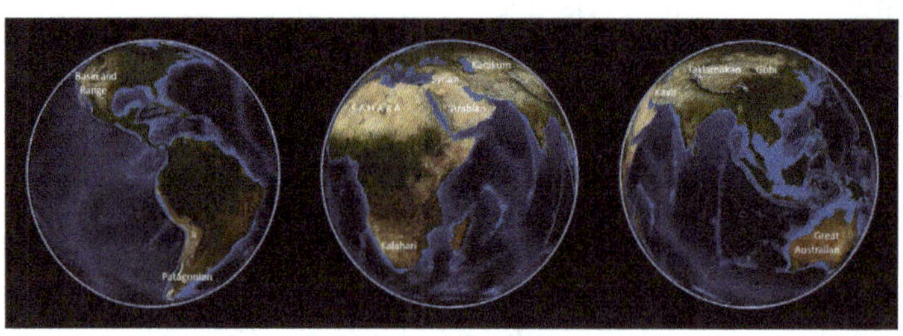

Largest deserts

The northern half of Africa is occupied by the world's most extensive hot, dry region, the Sahara Desert. Some deserts are also occupying much of southern Africa : the Namib and the Kalahari. Across Asia, a large annual rainfall minimum, composed primarily of deserts, stretches from the Gobi Desert in Mongolia west-southwest through western Pakistan (Balochistan) and Iran into the Arabian Desert in Saudi Arabia. Most of Australia is semi-arid or desert, making it the world's driest inhabited continent. In South America, the Andes mountain range blocks Pacific moisture that arrives in that continent, resulting in a desertlike climate just downwind across western Argentina. The drier areas of the United States are regions where the Sonoran Desert overspreads the Desert Southwest, the Great Basin and central Wyoming.

Isolated towering vertical desert shower

Polar Deserts

Since rain only falls as liquid, in frozen temperatures, rain cannot fall. As a result, very cold climates see very little rainfall and are often known as polar deserts. A common biome in this area is the tundra which has a short summer thaw and a long frozen winter. Ice caps see no rain at all, making Antarctica the world's driest continent.

Rainforests

Rainforests are areas of the world with very high rainfall. Both tropical and temperate rainforests exist. Tropical rainforests occupy a large band of the planet mostly along the equator. Most temperate rainforests are located on mountainous west coasts between 45 and 55 degrees latitude, but they are often found in other areas.

Around 40–75% of all biotic life is found in rainforests. Rainforests are also responsible for 28% of the world's oxygen turnover.

Monsoons

The equatorial region near the Intertropical Convergence Zone (ITCZ), or monsoon trough, is the wettest portion of the world's continents. Annually, the rain belt within the tropics marches northward by August, then moves back southward into the Southern Hemisphere by February and March. Within Asia, rainfall is favored across its southern portion from India east and northeast across the Philippines and southern China into Japan due to the monsoon advecting moisture primarily from the Indian Ocean into the region. The monsoon trough can reach as far north as the 40th parallel in East Asia during August before moving southward thereafter. Its poleward progression is accelerated by the onset of the summer monsoon which is characterized by the development of lower air pressure (a thermal low) over the warmest part of Asia. Similar, but weaker, monsoon circulations are present over North America and Australia. During the summer, the Southwest monsoon combined with Gulf of California and Gulf of Mexico moisture moving around the subtropical ridge in the Atlantic Ocean bring the promise of afternoon and evening thunderstorms to the southern tier of the United States as well as the Great Plains. The eastern half of the contiguous United States east of the

98th meridian, the mountains of the Pacific Northwest, and the Sierra Nevada range are the wetter portions of the nation, with average rainfall exceeding 760 mm (30 in) per year. Tropical cyclones enhance precipitation across southern sections of the United States, as well as Puerto Rico, the United States Virgin Islands, the Northern Mariana Islands, Guam, and American Samoa.

Impact of the Westerlies

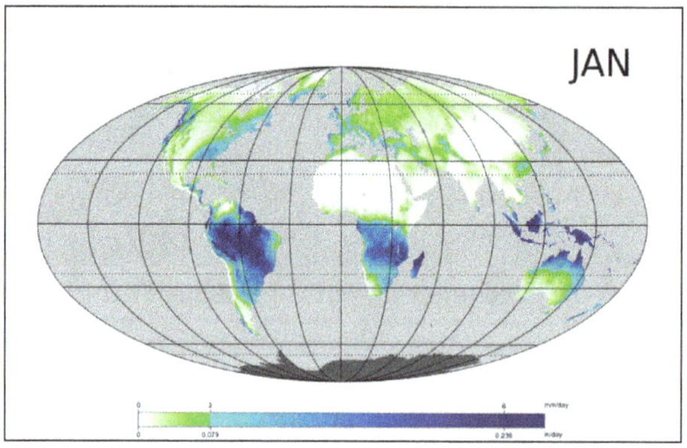

Long-term mean precipitation by month

Westerly flow from the mild north Atlantic leads to wetness across western Europe, in particular Ireland and the United Kingdom, where the western coasts can receive between 1,000 mm (39 in), at sea-level and 2,500 mm (98 in), on the mountains of rain per year. Bergen, Norway is one of the more famous European rain-cities with its yearly precipitation of 2,250 mm (89 in) on average. During the fall, winter, and spring, Pacific storm systems bring most of Hawaii and the western United States much of their precipitation. Over the top of the ridge, the jet stream brings a summer precipitation maximum to the Great Lakes. Large thunderstorm areas known as mesoscale convective complexes move through the Plains, Midwest, and Great Lakes during the warm season, contributing up to 10% of the annual precipitation to the region.

The El Niño-Southern Oscillation affects the precipitation distribution, by altering rainfall patterns across the western United States, Midwest, the Southeast, and throughout the tropics. There is also evidence that global warming is leading to increased precipitation to the eastern portions of North America, while droughts are becoming more frequent in the tropics and subtropics.

Wettest Known Locations

Cherrapunji, situated on the southern slopes of the Eastern Himalaya in Shillong, India is the confirmed wettest place on Earth, with an average annual rainfall of 11,430 mm (450 in). The highest recorded rainfall in a single year was 22,987 mm (905.0 in) in 1861. The 38-year average at nearby Mawsynram, Meghalaya, India is 11,873 mm (467.4 in). The wettest spot in Australia is Mount Bellenden Ker in the north-east of the country which records an average of 8,000 mm (310 in) per year, with over 12,200 mm (480.3 in) of rain recorded during 2000. The Big Bog on the island of Maui has the highest average annual rainfall in the Hawaiian Islands, at 10,300 mm (404 in). Mount Wai'ale'ale on the island of Kaua'i achieves similar torrential rains, while slightly lower than that of the Big Bog, at 9,500 mm (373 in) of rain per year over the last 32 years, with a record 17,340 mm (683 in) in 1982. Its summit is considered one of the rainiest spots on earth, with a reported 350 days of rain per year.

Lloró, a town situated in Chocó, Colombia, is probably the place with the largest rainfall in the world, averaging 13,300 mm (523.6 in) per year. The Department of Chocó is extraordinarily humid. Tutunendaó, a small town situated in the same department, is one of the wettest estimated places on Earth, averaging 11,394 mm (448.6 in) per year; in 1974 the town received 26,303 mm (86 ft 3.6 in), the largest annual rainfall measured in Colombia. Unlike Cherrapunji, which receives most of its rainfall between April and September, Tutunendaó receives rain almost uniformly distributed throughout the year. Quibdó, the capital of Chocó, receives the most rain in the world among cities with over 100,000 inhabitants: 9,000 mm (354 in) per year. Storms in Chocó can drop 500 mm (20 in) of rainfall in a day. This amount is more than what falls in many cities in a year's time.

Continent	Highest average		Place	Elevation		Years of record
	in	mm		ft	m	
South America	523.6	13,299	Lloró, Colombia (estimated)	520	158	29
Asia	467.4	11,872	Mawsynram, India	4,597	1,401	39
Africa	405.0	10,287	Debundscha, Cameroon	30	9.1	32
Oceania	404.3	10,269	Big Bog, Maui, Hawaii (USA)	5,148	1,569	30
South America	354.0	8,992	Quibdo, Colombia	120	36.6	16
Australia	340.0	8,636	Mount Bellenden Ker, Queensland	5,102	1,555	9
North America	256.0	6,502	Henderson Lake, British Columbia	12	3.66	14
Europe	183.0	4,648	Crkvice, Montenegro	3,337	1,017	22

	Continent	Place	Highest rainfall	
			in	mm
Highest average annual rainfall	Asia	Mawsynram, India	467.4	11,870
Highest in one year	Asia	Cherrapunji, India	1,042	26,470
Highest in one calendar month	Asia	Cherrapunji, India	366	9,296
Highest in 24 hours	Indian Ocean	Foc Foc, La Reunion Island	71.8	1,820
Highest in 12 hours	Indian Ocean	Foc Foc, La Reunion Island	45.0	1,140
Highest in one minute	North America	Unionville, Maryland, USA	1.23	31.2

FREEZING RAIN

Freezing rain is a rare type of liquid precipitation that strikes a cold surface, and freezes almost instantly.

The conditions needed for freezing rain are quite specific and we don't see this phenomenon very often in the UK. It can produce striking effects, as the rain drop spreads out momentarily across the surface before it freezes, encasing the surface in a layer of clear ice.

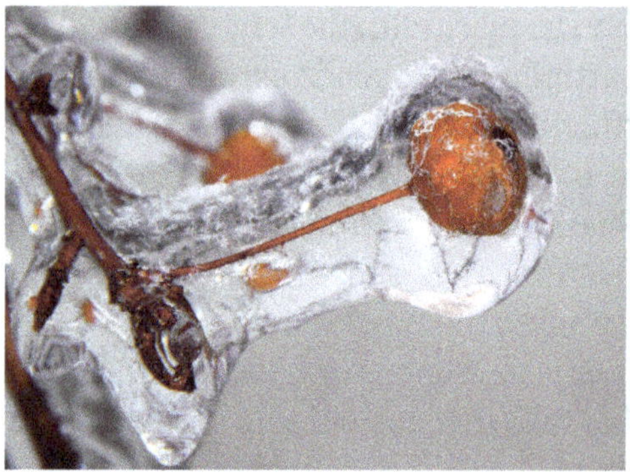

However, it is not just these eye-catching scenes which the freezing rain can bring; the weight of the ice can sometimes be heavy enough to bring down trees and power lines, and the glaze of ice on the ground effectively turns roads and pathways into an ice rink. The freezing rain can also prove extremely hazardous for aircraft.

Freezing rain is more common in other parts of the world, for example in the USA, where weather systems produce a lot of freezing rain. These are called ice storms, and if enough glaze collects on trees or power lines, the weight of the ice can cause them to break and can result in disruption on a large scale.

Formation of Freezing Rain

Very often, precipitation first falls from a cloud as snow (when it is cold enough high up where the cloud is). If it falls through warmer air before reaching the ground, it can melt and turn to rain droplets.

On rare occasions, if it then falls through cold air again just before hitting the ground, the droplets can become 'supercooled' and this means that they are still falling in liquid form, even though their temperature has fallen below zero. When this 'supercooled' droplet hits the ground (which is below zero too) it spreads out a little on landing, and then instantly freezes, encasing the surface in a layer of clear ice. This is why it is called freezing rain.

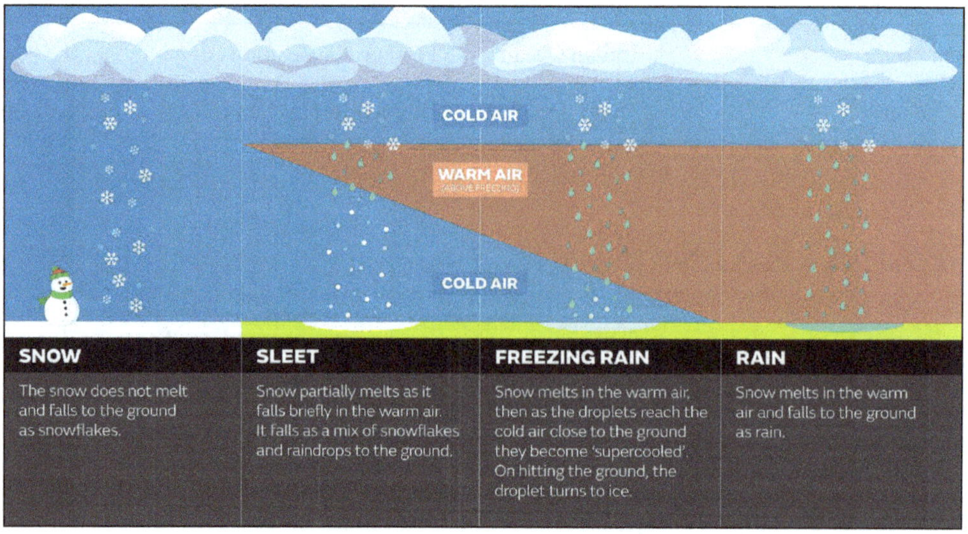

How does the Droplet become Supercooled?

We all know that water freezes at 0 °C but in the same way that raindrops need a nucleus (microscopic particles of dust or dirt) to form around, water droplets can only freeze around an ice nucleus. However, it is possible that water droplets can exist several degrees below zero and remain in liquid form without a nucleus. This is when they are supercooled.

Supercooled water droplets occur in clouds a lot of the time, especially in winter, and they form a very important role in the way that rain forms.

Freezing rain tends to start its life as snow, ice, sleet or hail, but passes through a layer of air that's above 0 °C on the way down to the ground, melting into a liquid water droplet. If these droplets then fall through a zone of sub-zero air just above the ground, they become supercooled. When these supercooled droplets strike surfaces that are close to or below freezing, they freeze on impact forming a glaze of ice.

Effects

Freezing Rain can cause significant accumulations of clear ice on upper surfaces which the aircraft anti/de-icing systems, being primarily designed to deal with ice accumulation caused by horizontal movement of the airframe, may not be able to cope with. Aircraft on the ground during freezing rain may become covered in thick layers of ice which will be difficult to remove; airfield operating surfaces and runways will also be affected, and all aircraft movement on the ground may be severely disrupted.

Solutions

- In Flight: The best response is avoidance: aircraft should be routed away from forecast/reported freezing rain horizontally and/or vertically. When in the vicinity of a warm front in winter, if performance, ATC, and regulations allow, consider flying above the freezing level, where the ice crystal precipitation will not have a chance to melt.
- On the Ground: Check information on runway braking action before taxiing for take-off or attempting a take-off or landing. Generally airline and passenger/freight carrying operations are suspended during periods of freezing rain. Even without being prohibited by the operators manual, operations in freezing rain are difficult as the applicable Hold-Over Time from an anti-icing treatment is very limiting in freezing rain. This is because the ice accretion rates due to freezing rain can be much greater than those associated with other forms of winter precipitation. A typical hold over time can be as little as 15 minutes in light freezing rain, which can often mean that it is not possible to receive treatment and then take off within the allowable time frame. It should be noted that Holdover Time (HOT) Tables are not provided for operations in conditions of moderate or heavy freezing rain as the anti-icing fluid is not capable of providing protection under those circumstances. Ice pellets at surface level may indicate freezing rain aloft.

ATMOSPHERIC PRESSURE

Atmospheric pressure, sometimes also called barometric pressure (after the sensor), is the pressure within the atmosphere of Earth. The standard atmosphere (symbol: atm) is a unit of pressure defined as 1,013.25 mbar (101,325 Pa; 1,013.25 hPa), equivalent to 760 mm Hg (torr), 29.9212 inches Hg, or 14.696 psi. The atm unit is roughly equivalent to the mean sea-level atmospheric pressure on Earth, that is, the Earth's atmospheric pressure at sea level is approximately 1 atm.

In most circumstances atmospheric pressure is closely approximated by the hydrostatic pressure caused by the weight of air above the measurement point. As elevation increases, there is less overlying atmospheric mass, so that atmospheric pressure decreases with increasing elevation. Pressure measures force per unit area, with SI units of Pascals (1 pascal = 1 newton per square metre, 1 N/m^2). On average, a column of air with a cross-sectional area of 1 square centimetre (cm^2), measured from mean (average) sea level to the top of Earth's atmosphere, has a mass of about 1.03 kilogram and exerts a force or "weight" of about 10.1 newtons, resulting in a pressure of 10.1 N/cm^2 or 101 kN/m^2 (101 kilopascals, kPa). A column of air with a cross-sectional area of 1 in^2 would have a weight of about 14.7 lb_f, resulting in a pressure of 14.7 lb_f/in^2.

Mechanism

Atmospheric pressure is caused by the gravitational attraction of the planet on the atmospheric gases above the surface, and is a function of the mass of the planet, the radius of the surface, and the amount and composition of the gases and their vertical distribution in the atmosphere. It is modified by the planetary rotation and local effects such as wind velocity, density variations due to temperature and variations in composition.

Mean Sea-level Pressure

The *mean sea-level pressure* (MSLP) is the average atmospheric pressure at mean sea level. This is the atmospheric pressure normally given in weather reports on radio, television, and newspapers or on the Internet. When barometers in the home are set to match the local weather reports, they measure pressure adjusted to sea level, not the actual local atmospheric pressure.

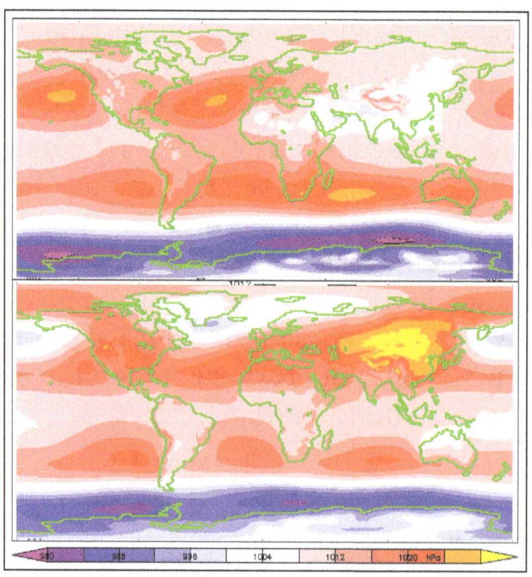

15-year average mean sea-level pressure for June, July, and August (top) and December, January, and February (bottom). ERA-15 re-analysis

Kollsman-type barometric aircraft altimeter (as used in North America) displaying an altitude of 80 ft (24 m)

The *altimeter setting* in aviation is an atmospheric pressure adjustment.

Average *sea-level pressure* is 1013.25 mbar (101.325 kPa; 29.921 inHg; 760.00 mmHg). In aviation, weather reports (METAR), QNH is transmitted around the world in millibars or hectopascals (1 hectopascal = 1 millibar), except in the United States, Canada, and Colombia where it is reported in inches of mercury (to two decimal places). The United States and Canada also report *sea-level pressure* SLP, which is adjusted to sea level by a different method, in the remarks section, not in the internationally transmitted part of the code, in hectopascals or millibars. However, in Canada's public weather reports, sea level pressure is instead reported in kilopascals.

In the US weather code remarks, three digits are all that are transmitted; decimal points and the one or two most significant digits are omitted: 1013.2 mbar (101.32 kPa) is transmitted as 132; 1000.0 mbar (100.00 kPa) is transmitted as 000; 998.7 mbar is transmitted as 987; etc. The highest *sea-level pressure* on Earth occurs in Siberia, where the Siberian High often attains a *sea-level pressure* above 1050 mbar (105 kPa; 31 inHg), with record highs close to 1085 mbar (108.5 kPa; 32.0 inHg). The lowest measurable *sea-level pressure* is found at the centers of tropical cyclones and tornadoes, with a record low of 870 mbar (87 kPa; 26 inHg).

Surface Pressure

Surface pressure is the atmospheric pressure at a location on Earth's surface (terrain and oceans). It is directly proportional to the mass of air over that location.

For numerical reasons, atmospheric models such as general circulation models (GCMs) usually predict the nondimensional logarithm of surface pressure.

The average value of surface pressure on Earth is 985 hPa. This is in contrast to mean sea-level pressure, which involves the extrapolation of pressure to sea-level for locations above or below sea-level. The average pressure at mean sea-level (MSL) in the International Standard Atmosphere (ISA) is 1013.25 hPa, or 1 atmosphere (Atm), or 29.92 inches of mercury.

Pressure (P), mass (m), and the acceleration due to gravity (g), are related by P = F/A = (m*g)/A, where A is surface area. Atmospheric pressure is thus proportional to the weight per unit area of the atmospheric mass above that location.

Altitude Variation

A very local storm above Snæfellsjökull, showing clouds formed on the mountain by orographic lift

Pressure on Earth varies with the altitude of the surface; so air pressure on mountains is usually lower than air pressure at sea level. Pressure varies smoothly from the Earth's surface to the top of the mesosphere. Although the pressure changes with the weather, NASA has averaged the conditions for all parts of the earth year-round. As altitude increases, atmospheric pressure decreases. One can calculate the atmospheric pressure at a given altitude. Temperature and humidity also affect the atmospheric pressure, and it is necessary to know these to compute an accurate figure. The graph was developed for a temperature of 15 °C and a relative humidity of 0%.

This plastic bottle was sealed at approximately 14,000 feet (4,300 m) altitude, and was crushed by the increase in atmospheric pressure, recorded at 9,000 feet (2,700 m) and 1,000 feet (300 m), as it was brought down towards sea level

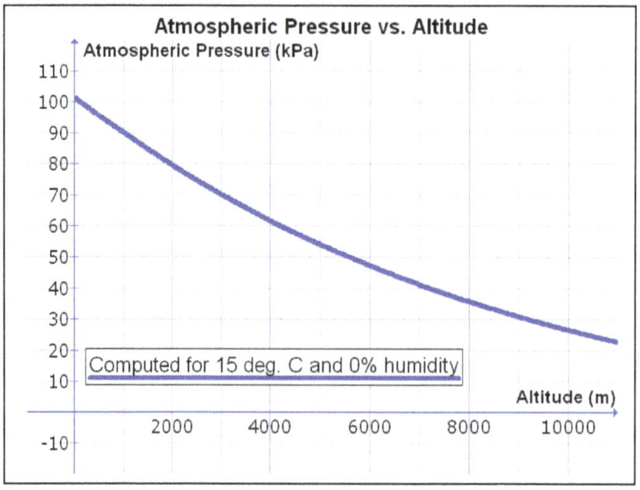

Variation in atmospheric pressure with altitude, computed for 15 °C and 0% relative humidity

At low altitudes above sea level, the pressure decreases by about 1.2 kPa (12 hPa) for every 100 metres. For higher altitudes within the troposphere, the following equation (the barometric formula) relates atmospheric pressure p to altitude h:

$$p = p_0 \cdot \left(1 - \frac{L \cdot h}{T_0}\right)^{\frac{g \cdot M}{R_0 \cdot L}}$$

$$= p_0 \cdot \left(1 - \frac{g \cdot h}{c_p \cdot T_0}\right)^{\frac{c_p \cdot M}{R_0}} \approx p_0 \cdot \exp\left(-\frac{g \cdot h \cdot M}{T_0 \cdot R_0}\right)$$

where the constant parameters are as described below:

Parameter	Description	Value
p_o	Sea level standard atmospheric pressure	101325 Pa
L	Temperature lapse rate, = g/c_p for dry air	~ 0.00976 K/m
c_p	Constant-pressure specific heat	1004.68506 J/(kg·K)
T_o	Sea level standard temperature	288.16 K
g	Earth-surface gravitational acceleration	9.80665 m/s²
M	Molar mass of dry air	0.02896968 kg/mol
R_o	Universal gas constant	8.314462618 J/(mol·K)

Local Variation

Atmospheric pressure varies widely on Earth, and these changes are important in studying weather and climate.

Atmospheric pressure shows a diurnal or semidiurnal (twice-daily) cycle caused by global atmospheric tides. This effect is strongest in tropical zones, with an amplitude of a few millibars, and almost zero in polar areas. These variations have two superimposed cycles, a circadian (24 h) cycle and semi-circadian (12 h) cycle.

Records

The highest adjusted-to-sea level barometric pressure ever recorded on Earth (above 750 meters) was 1084.8 hPa (32.03 inHg) measured in Tosontsengel, Mongolia on 19 December 2001. The highest adjusted-to-sea level barometric pressure ever recorded (below 750 meters) was at Agata in Evenk Autonomous Okrug, Russia (66°53' N, 93°28' E, elevation: 261 m, 856 ft) on 31 December 1968 of 1083.8 hPa (32.005 inHg). The discrimination is due to the problematic assumptions (assuming a standard lapse rate) associated with reduction of sea level from high elevations.

The Dead Sea, the lowest place on Earth at 430 metres (1,410 ft) below sea level, has a correspondingly high typical atmospheric pressure of 1065 hPa.

Hurricane Wilma on 19 October 2005; 882 hPa (12.79 psi) in the storm's eye

The lowest non-tornadic atmospheric pressure ever measured was 870 hPa (0.858 atm; 25.69 inHg), set on 12 October 1979, during Typhoon Tip in the western Pacific Ocean. The measurement was based on an instrumental observation made from a reconnaissance aircraft.

Measurement Based on Depth of Water

One atmosphere (101.325 kPa or 14.7 psi) is also the pressure caused by the weight of a column of fresh water of approximately 10.3 m (33.8 ft). Thus, a diver 10.3 m underwater experiences a pressure of about 2 atmospheres (1 atm of air plus 1 atm of water). Conversely, 10.3 m is the maximum height to which water can be raised using suction under standard atmospheric conditions.

Low pressures such as natural gas lines are sometimes specified in inches of water, typically written as *w.c.* (water column) gauge or *w.g.* (inches water gauge). A typical gas-using residential appliance in the US is rated for a maximum of 1/2 psi, which is approximately 14 w.g. (3487 Pa or 34.9 millibars). Similar metric units with a wide variety of names and notation based on millimetres, centimetres or metres are now less commonly used.

Boiling Point of Water

Boiling water

Pure water boils at 100 °C (212 °F) at earth's standard atmospheric pressure. The boiling point is the temperature at which the vapor pressure is equal to the atmospheric pressure around the water. Because of this, the boiling point of water is lower at lower pressure and higher at higher pressure. Cooking at high elevations, therefore, requires adjustments to recipes or pressure cooking. A rough approximation of elevation can be obtained by measuring the temperature at which water boils; in the mid-19th century, this method was used by explorers.

Measurement and Maps

An important application of the knowledge that atmospheric pressure varies directly with altitude was in determining the height of hills and mountains thanks to the availability of reliable pressure measurement devices. In 1774, Maskelyne was confirming Newton's theory of gravitation at and on Schiehallion mountain in Scotland, and he needed to accurately measure elevations on the mountain's sides. William Roy, using barometric pressure, was able to confirm Maskelyne's height determinations, the agreement being to within one meter (3.28 feet). This method became and continues to be useful for survey work and map making.

WATER VAPOR

Water vapor, water vapour or aqueous vapor is the gaseous phase of water. It is one state of water within the hydrosphere. Water vapor can be produced from the evaporation or boiling of liquid water or from the sublimation of ice. Unlike other forms of water, water vapor is invisible. Under typical atmospheric conditions, water vapor is continuously generated by evaporation and removed by condensation. It is less dense than air and triggers convection currents that can lead to clouds.

Being a component of Earth's hydrosphere and hydrologic cycle, it is particularly abundant in Earth's atmosphere where it is also a potent greenhouse gas along with other gases such as carbon dioxide and methane. Use of water vapor, as steam, has been important to humans for cooking and as a major component in energy production and transport systems since the industrial revolution.

Water vapor is a relatively common atmospheric constituent, present even in the solar atmosphere as well as every planet in the Solar System and many astronomical objects including natural satellites, comets and even large asteroids. Likewise the detection of extrasolar water vapor would indicate a similar distribution in other planetary systems. Water vapor is significant in that it can be indirect evidence supporting the presence of extraterrestrial liquid water in the case of some planetary mass objects.

Properties

Evaporation

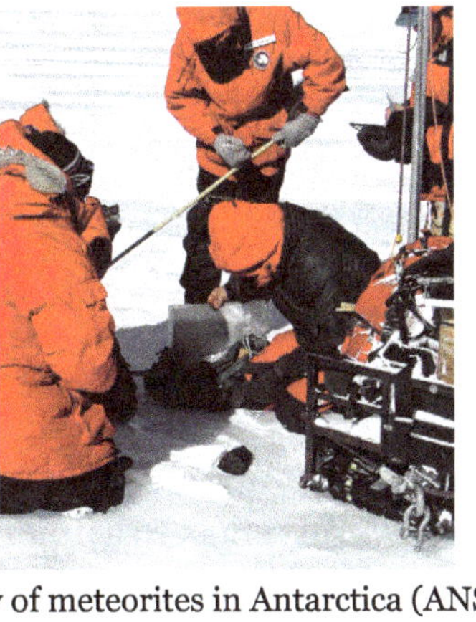

Recovery of meteorites in Antarctica (ANSMET)

Whenever a water molecule leaves a surface and diffuses into a surrounding gas, it is said to have evaporated. Each individual water molecule which transitions between a more associated (liquid) and a less associated (vapor/gas) state does so through the absorption or release of kinetic energy. The aggregate measurement of this kinetic energy transfer is defined as thermal energy and occurs only when there is differential in the temperature of the water molecules. Liquid water that becomes water vapor takes a parcel of heat with it, in a process called evaporative cooling. The amount of water vapor in the air determines how frequently molecules will return to the surface. When a net evaporation occurs, the body of water will undergo a net cooling directly related to the loss of water.

In the US, the National Weather Service measures the actual rate of evaporation from a standardized "pan" open water surface outdoors, at various locations nationwide. Others do likewise around the world. The US data is collected and compiled into an annual evaporation map. The measurements range from under 30 to over 120 inches per year. Formulas can be used for calculating the rate of evaporation from a water surface such as a swimming pool. In some countries, the evaporation rate far exceeds the precipitation rate.

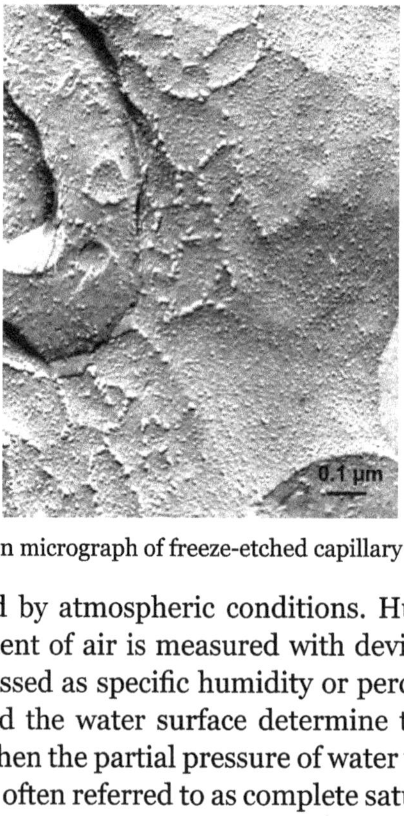

Electron micrograph of freeze-etched capillary tissue

Evaporative cooling is restricted by atmospheric conditions. Humidity is the amount of water vapor in the air. The vapor content of air is measured with devices known as hygrometers. The measurements are usually expressed as specific humidity or percent relative humidity. The temperatures of the atmosphere and the water surface determine the equilibrium vapor pressure; 100% relative humidity occurs when the partial pressure of water vapor is equal to the equilibrium vapor pressure. This condition is often referred to as complete saturation. Humidity ranges from 0 gram per cubic metre in dry air to 30 grams per cubic metre (0.03 ounce per cubic foot) when the vapor is saturated at 30 °C.

Sublimation

Sublimation is when water molecules directly leave the surface of ice without first becoming liquid water. Sublimation accounts for the slow mid-winter disappearance of ice and snow at temperatures too low to cause melting. Antarctica shows this effect to a unique degree because it is by far the continent with the lowest rate of precipitation on Earth. As a result, there are large areas where millennial layers of snow have sublimed, leaving behind whatever non-volatile materials they had contained. This is extremely valuable to certain scientific disciplines, a dramatic example being the collection of meteorites that are left exposed in unparalleled numbers and excellent states of preservation.

Sublimation is important in the preparation of certain classes of biological specimens for scanning electron microscopy. Typically the specimens are prepared by cryofixation and freeze-fracture, after which the broken surface is freeze-etched, being eroded by exposure to vacuum till it shows the required level of detail. This technique can display protein molecules, organelle structures and lipid bilayers with very low degrees of distortion.

Condensation

Water vapour will only condense onto another surface when that surface is cooler than the dew point temperature, or when the water vapour equilibrium in air has been exceeded. When water vapour condenses onto a surface, a net warming occurs on that surface. The water molecule brings heat energy with it. In turn, the temperature of the atmosphere drops slightly. In the atmosphere, condensation produces clouds, fog and precipitation (usually only when facilitated by cloud condensation nuclei). The dew point of an air parcel is the temperature to which it must cool before water vapour in the air begins to condense concluding water vapour is a type of water or rain.

Clouds, formed by condensed water vapour

Also, a net condensation of water vapour occurs on surfaces when the temperature of the surface is at or below the dew point temperature of the atmosphere. Deposition is a phase transition separate from condensation which leads to the direct formation of ice from water vapour. Frost and snow are examples of deposition.

There are many mechanism of cooling by which condensation occurs 1.direct loss of heat known as radiation cooling. 2.cooling with upliftment of air known as adiabatic cooling. there are 4 types of upliftment of air: a. orographic upliftment- mountains work as barrier for upliftment of air b. convectional upliftment- upliftment of air due to pressure difference c. frontal upliftment- upliftment of air due to temp difference d. cyclonic upliftment but not related to temp 3.advective cooling-cooling due to horizontal movement of air.

Chemical Reactions

A number of chemical reactions have water as a product. If the reactions take place at temperatures higher than the dew point of the surrounding air the water will be formed as vapor and increase the local humidity, if below the dew point local condensation will occur. Typical reactions that result in water formation are the burning of hydrogen or hydrocarbons in air or other oxygen containing gas mixtures, or as a result of reactions with oxidizers.

In a similar fashion other chemical or physical reactions can take place in the presence of water vapor resulting in new chemicals forming such as rust on iron or steel, polymerization occurring (certain polyurethane foams and cyanoacrylate glues cure with exposure to atmospheric humidity) or forms changing such as where anhydrous chemicals may absorb enough vapor to form a crystalline structure or alter an existing one, sometimes resulting in characteristic color changes that can be used for measurement.

Measurement

Measuring the quantity of water vapor in a medium can be done directly or remotely with varying degrees of accuracy. Remote methods such electromagnetic absorption are possible from satellites above planetary atmospheres. Direct methods may use electronic transducers, moistened thermometers or hygroscopic materials measuring changes in physical properties or dimensions.

	Medium	Temperature range (degC)	Measurement uncertainty	Typical Measurement Frequency	System Cost
Sling psychrometer	Air	−10 to 50	Low to moderate	Hourly	Low
Satellite-based spectroscopy	Air	−80 to 60	Low		Very high
Capacitive sensor	Air/gases	−40 to 50	Moderate	2 to 0.05 Hz	Medium
Warmed capacitive sensor	Air/gases	−15 to 50	Moderate to low	2 to 0.05 Hz (Temp dependant)	Medium to high
Resistive sensor	Air/gases	−10 to 50	Moderate	60 seconds	Medium
Lithium chloride dewcell	Air	−30 to 50	Moderate	Continuous	Medium
Cobalt(II) chloride	Air/gases	0 to 50	High	5 minutes	Very low
Absorption spectroscopy	Air/gases		Moderate		High
Aluminum oxide	Air/gases		Moderate		Medium
Silicon oxide	Air/gases		Moderate		Medium
Piezoelectric sorption	Air/gases		Moderate		Medium
Electrolytic	Air/gases		Moderate		Medium
Hair tension	Air	0 to 40	High	Continuous	Low to medium
Nephelometer	Air/other gases		Low		Very high
Goldbeater's skin (Cow Peritoneum)	Air	−20 to 30	Moderate (with corrections)	Slow, slower at lower temperatures	Low
Lyman-alpha				High frequency	High
Gravimetric Hygrometer			Very low		Very high
	Medium	Temperature range (degC)	Measurement uncertainty	Typical measurement frequency	System cost

Impact on Air Density

Water vapor is lighter or less dense than dry air. At equivalent temperatures it is buoyant with respect to dry air, whereby the density of dry air at standard temperature and pressure (273.15 K, 101.325 kPa) is 1.27 g/L and water vapor at standard temperature has a vapor pressure of 0.6 kPa and the much lower density of 4.85 mg/L.

Calculations

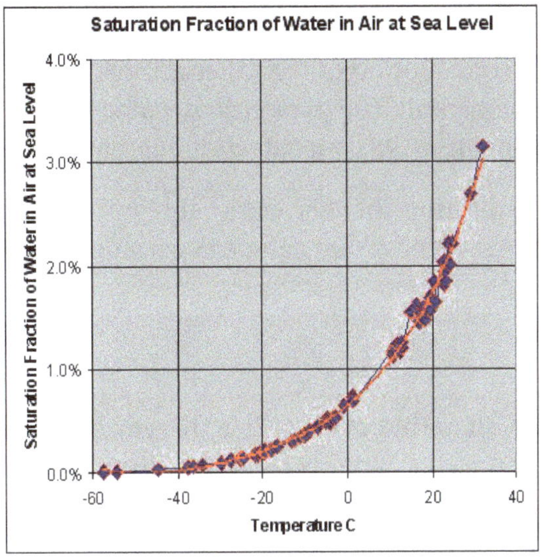

Water vapor and dry air density calculations at 0 °C:

- The molar mass of water is 18.02 g/mol, as calculated from the sum of the atomic masses of its constituent atoms.
- The average molecular mass of air (approx. 78% nitrogen, N_2; 21% oxygen, O_2; 1% other gases) is 28.57 g/mol at standard temperature and pressure (STP).
- Obeying Avogadro's Law and the ideal gas law, moist air will have a lower density than dry air. At max. saturation (i. e. rel. humidity = 100 % at 0 °C) the density will go down to 28.51 g/mol.
- STP conditions imply a temperature of 0 °C, at which the ability of water to become vapor is very restricted. Its concentration in air is very low at 0 °C. The red line on the chart to the right is the maximum concentration of water vapor expected for a given temperature. The water vapor concentration increases significantly as the temperature rises, approaching 100% (steam, pure water vapor) at 100 °C. However the difference in densities between air and water vapor would still exist (0.598 vs. 1.27 g/l).

At Equal Temperatures

At the same temperature, a column of dry air will be denser or heavier than a column of air containing any water vapor, the molar mass of diatomic nitrogen and diatomic oxygen both being greater than the molar mass of water. Thus, any volume of dry air will sink if placed in a larger volume of moist air. Also, a volume of moist air will rise or be buoyant if placed in a larger region of dry air.

As the temperature rises the proportion of water vapor in the air increases, and its buoyancy will increase. The increase in buoyancy can have a significant atmospheric impact, giving rise to powerful, moisture rich, upward air currents when the air temperature and sea temperature reaches 25 °C or above. This phenomenon provides a significant driving force for cyclonic and anticyclonic weather systems (typhoons and hurricanes).

Respiration and Breathing

Water vapor is a by-product of respiration in plants and animals. Its contribution to the pressure, increases as its concentration increases. Its partial pressure contribution to air pressure increases, lowering the partial pressure contribution of the other atmospheric gases (Dalton's Law). The total air pressure must remain constant. The presence of water vapor in the air naturally dilutes or displaces the other air components as its concentration increases.

This can have an effect on respiration. In very warm air (35 °C) the proportion of water vapor is large enough to give rise to the stuffiness that can be experienced in humid jungle conditions or in poorly ventilated buildings.

Lifting Gas

Water vapor has lower density than that of air and is therefore buoyant in air but has lower vapor pressure than that of air. When water vapor is used as a lifting gas by a thermal airship the water vapor is heated to form steam so that its vapor pressure is greater than the surrounding air pressure in order to maintain the shape of a theoretical "steam balloon", which yields approximately 60% the lift of helium and twice that of hot air.

The amount of water vapor in an atmosphere is constrained by the restrictions of partial pressures and temperature. Dew point temperature and relative humidity act as guidelines for the process of water vapor in the water cycle. Energy input, such as sunlight, can trigger more evaporation on an ocean surface or more sublimation on a chunk of ice on top of a mountain. The *balance* between condensation and evaporation gives the quantity called vapor partial pressure.

The maximum partial pressure (*saturation pressure*) of water vapor in air varies with temperature of the air and water vapor mixture. A variety of empirical formulas exist for this quantity; the most used reference formula is the Goff-Gratch equation for the SVP over liquid water below zero degree Celsius:

$$\begin{aligned}\log_{10}(p) = & -7.90298\left(\frac{373.16}{T}-1\right)+5.02808\log_{10}\frac{373.16}{T} \\ & -1.3816\times10^{-7}\left(10^{11.344\left(1-\frac{T}{373.16}\right)}-1\right) \\ & +8.1328\times10^{-3}\left(10^{-3.49149\left(\frac{373.16}{T}-1\right)}-1\right) \\ & +\log_{10}(1013.246)\end{aligned}$$

Where T, temperature of the moist air, is given in units of kelvin, and p is given in units of millibars (hectopascals).

The formula is valid from about −50 to 102 °C; however there are a very limited number of measurements of the vapor pressure of water over supercooled liquid water. There are a number of other formulae which can be used.

Under certain conditions, such as when the boiling temperature of water is reached, a net evaporation will always occur during standard atmospheric conditions regardless of the percent of relative humidity. This immediate process will dispel massive amounts of water vapor into a cooler atmosphere.

Exhaled air is almost fully at equilibrium with water vapor at the body temperature. In the cold air the exhaled vapor quickly condenses, thus showing up as a fog or mist of water droplets and as condensation or frost on surfaces. Forcibly condensing these water droplets from exhaled breath is the basis of exhaled breath condensate, an evolving medical diagnostic test.

Controlling water vapor in air is a key concern in the heating, ventilating, and air-conditioning (HVAC) industry. Thermal comfort depends on the moist air conditions. Non-human comfort situations are called refrigeration, and also are affected by water vapor. For example, many food stores, like supermarkets, utilize open chiller cabinets, or *food cases*, which can significantly lower the water vapor pressure (lowering humidity). This practice delivers several benefits as well as problems.

In Earth's Atmosphere

Gaseous water represents a small but environmentally significant constituent of the atmosphere. The percentage water vapor in surface air varies from 0.01% at -42 °C (-44 °F) to 4.24% when the dew point is 30 °C (86 °F). Approximately 99.13% of it is contained in the troposphere. The condensation of water vapor to the liquid or ice phase is responsible for clouds, rain, snow, and other precipitation, all of which count among the most significant elements of what we experience as weather. Less obviously, the latent heat of vaporization, which is released to the atmosphere whenever condensation occurs, is one of the most important terms in the atmospheric energy budget on both local and global scales. For example, latent heat release in atmospheric convection is directly responsible for powering destructive storms such as tropical cyclones and severe thunderstorms. Water vapor is the most potent greenhouse gas owing to the presence of the hydroxyl bond which strongly absorbs in the infra-red region of the light spectrum.

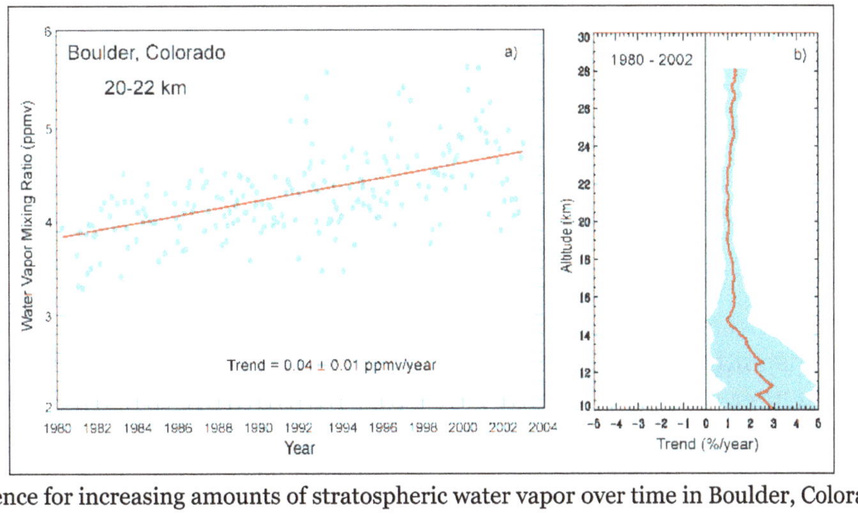

Evidence for increasing amounts of stratospheric water vapor over time in Boulder, Colorado

Water vapor is the "working medium" of the atmospheric thermodynamic engine which transforms heat energy from sun irradiation into mechanical energy in the form of winds. Transforming thermal energy into mechanical energy requires an upper and a lower temperature level, as well as a working medium which shuttles forth and back between both. The upper temperature level is given by the soil or water surface of the earth, which absorbs the incoming sun radiation and warms up, evaporating water. The moist and warm air at the ground is lighter than its surrounding and raises up to the upper limit of the troposphere. There the water molecules radiate their thermal energy into outer space, cooling down the surrounding air. The upper atmosphere constitutes the lower temperature level of the atmospheric thermodynamic engine. The water vapor in the now cold air condenses out and falls down to the ground in the form of rain or snow. The now heavier cold and dry air sinks down to ground as well; the atmospheric thermodynamic engine thus establishes a vertical convection, which transports heat from the ground into the upper atmosphere, where the water molecules can radiate it to outer space. Due to the earth rotation and the resulting Coriolis forces, this vertical atmospheric convection is also converted into a horizontal convection, in the form of cyclones and anticyclones, which transport the water evaporated over the oceans into the interior of the continents, enabling vegetation to grow.

Water in Earth's atmosphere is not merely below its boiling point (100 °C), but at altitude it goes below its freezing point (0 °C), due to water's highly polar attraction. When combined with its quantity, water vapor then has a relevant dew point and frost point, unlike e. g., carbon dioxide and methane. Water vapor thus has a scale height a fraction of that of the bulk atmosphere, as the water condenses and exits, primarily in the troposphere, the lowest layer of the atmosphere. Carbon dioxide (CO_2) and methane, being non-polar, rise above water vapor. The absorption and emission of both compounds contribute to Earth's emission to space, and thus the planetary greenhouse effect. This greenhouse forcing is directly observable, via distinct spectral features versus water vapor, and observed to be rising with rising CO_2 levels. Conversely, adding water vapor at high altitudes has a disproportionate impact, which is why methane (rising, then oxidizing to CO_2 and two water molecules) and jet traffic have disproportionately high warming effects.

It is less clear how cloudiness would respond to a warming climate; depending on the nature of the response, clouds could either further amplify or partly mitigate warming from long-lived greenhouse gases.

In the absence of other greenhouse gases, Earth's water vapor would condense to the surface; this has likely happened, possibly more than once. Scientists thus distinguish between non-condensable (driving) and condensable (driven) greenhouse gases.

Fog and clouds form through condensation around cloud condensation nuclei. In the absence of nuclei, condensation will only occur at much lower temperatures. Under persistent condensation or deposition, cloud droplets or snowflakes form, which precipitate when they reach a critical mass.

The water content of the atmosphere as a whole is constantly depleted by precipitation. At the same time it is constantly replenished by evaporation, most prominently from seas, lakes, rivers, and moist earth. Other sources of atmospheric water include combustion, respiration, volcanic eruptions, the transpiration of plants, and various other biological and geological processes. The mean global content of water vapor in the atmosphere is roughly sufficient to cover the surface of the planet with a layer of liquid water about 25 mm deep. The mean annual precipitation for the planet is about 1 meter, which implies a rapid turnover of water in the air – on average, the residence time of a water molecule in the troposphere is about 9 to 10 days.

Episodes of surface geothermal activity, such as volcanic eruptions and geysers, release variable amounts of water vapor into the atmosphere. Such eruptions may be large in human terms, and major explosive eruptions may inject exceptionally large masses of water exceptionally high into the atmosphere, but as a percentage of total atmospheric water, the role of such processes is trivial. The relative concentrations of the various gases emitted by volcanoes varies considerably according to the site and according to the particular event at any one site. However, water vapor is consistently the commonest volcanic gas; as a rule, it comprises more than 60% of total emissions during a subaerial eruption.

Atmospheric water vapor content is expressed using various measures. These include vapor pressure, specific humidity, mixing ratio, dew point temperature, and relative humidity.

Radar and Satellite Imaging

Because water molecules absorb microwaves and other radio wave frequencies, water in the atmosphere attenuates radar signals. In addition, atmospheric water will reflect and refract signals to an extent that depends on whether it is vapor, liquid or solid.

Generally, radar signals lose strength progressively the farther they travel through the troposphere. Different frequencies attenuate at different rates, such that some components of air are opaque to some frequencies and transparent to others. Radio waves used for broadcasting and other communication experience the same effect.

These maps show the average amount of water vapor in a column of atmosphere in a given month

Water vapor reflects radar to a lesser extent than do water's other two phases. In the form of drops and ice crystals, water acts as a prism, which it does not do as an individual molecule; however, the existence of water vapor in the atmosphere causes the atmosphere to act as a giant prism.

A comparison of GOES-12 satellite images shows the distribution of atmospheric water vapor relative to the oceans, clouds and continents of the Earth. Vapor surrounds the planet but is unevenly distributed. The image shows monthly average of water vapor content with the units are given in centimeters, which is the precipitable water or equivalent amount of water that could be produced if all the water vapor in the column were to condense. The lowest amounts of water vapor (0 centimeters) appear in yellow, and the highest amounts (6 centimeters) appear in dark blue. Areas of missing data appear in shades of gray. The most noticeable pattern in the time series is the influence of seasonal temperature changes and incoming sunlight on water vapor. In the tropics, a band of extremely humid air wobbles north and south of the equator as the seasons change. This band of humidity is part of the Intertropical Convergence Zone, where the easterly trade winds from each hemisphere converge and produce near-daily thunderstorms and clouds. Farther from the equator, water vapor concentrations are high in the hemisphere experiencing summer and low in the one experiencing winter. Another pattern that shows up in the time series is that water vapor amounts over land areas decrease more in winter months than adjacent ocean areas do. This is largely because air temperatures over land drop more in the winter than temperatures over the ocean. Water vapor condenses more rapidly in colder air.

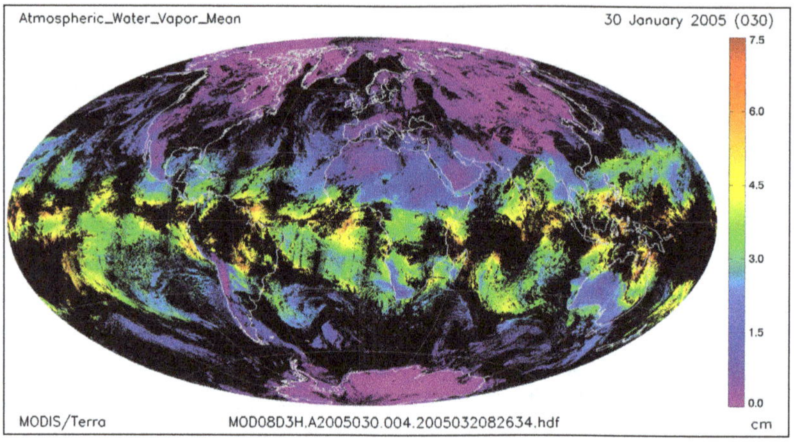

MODIS/Terra global mean atmospheric water vapor in atm-cm (centimeters of water in an atmospheric column if it condensed)

As water vapour absorbs light in the visible spectral range, its absorption can be used in spectroscopic applications (such as DOAS) to determine the amount of water vapor in the atmosphere. This is done operationally, e.g. from the GOME spectrometers on ERS and MetOp. The weaker water vapor absorption lines in the blue spectral range and further into the UV up to its dissociation limit around 243 nm are mostly based on quantum mechanical calculations and are only partly confirmed by experiments.

Lightning Generation

Water vapor plays a key role in lightning production in the atmosphere. From cloud physics, usually clouds are the real generators of static charge as found in Earth's atmosphere. The ability of clouds to hold massive amounts of electrical energy is directly related to the amount of water vapor present in the local system.

The amount of water vapor directly controls the permittivity of the air. During times of low humidity, static discharge is quick and easy. During times of higher humidity, fewer static discharges occur. Permittivity and capacitance work hand in hand to produce the megawatt outputs of lightning.

After a cloud, for instance, has started its way to becoming a lightning generator, atmospheric water vapor acts as a substance (or insulator) that decreases the ability of the cloud to discharge its electrical energy. Over a certain amount of time, if the cloud continues to generate and store more static electricity, the barrier that was created by the atmospheric water vapor will ultimately break down from the stored electrical potential energy. This energy will be released to a local oppositely charged region, in the form of lightning. The strength of each discharge is directly related to the atmospheric permittivity, capacitance, and the source's charge generating ability.

WIND

Cherry tree moving with wind blowing about 22 m/sec (about 49 mph)

Wind is the flow of gases on a large scale. On the surface of the Earth, wind consists of the bulk movement of air. In outer space, solar wind is the movement of gases or charged particles from the Sun through space, while planetary wind is the outgassing of light chemical elements from a planet's atmosphere into space. Winds are commonly classified by their spatial scale, their speed, the types of forces that cause them, the regions in which they occur, and their effect. The strongest observed winds on a planet in the Solar System occur on Neptune and Saturn. Winds have various aspects: velocity (wind speed); the density of the gas involved; energy content or wind energy. Wind is also an important means of transportation for seeds and small birds; with time things can travel thousands of miles in the wind.

In meteorology, winds are often referred to according to their strength, and the direction from which the wind is blowing. Short bursts of high-speed wind are termed gusts. Strong winds of intermediate duration (around one minute) are termed squalls. Long-duration winds have various names associated with their average strength, such as breeze, gale, storm, and hurricane. Wind occurs on a range of scales, from thunderstorm flows lasting tens of minutes, to local breezes generated by heating of land surfaces and lasting a few hours, to global winds resulting from the difference in absorption of solar energy between the climate zones on Earth. The two main causes of large-scale atmospheric circulation are the differential heating between the equator and the poles, and the rotation of the planet (Coriolis effect). Within the tropics, thermal low circulations over terrain and high plateaus can drive monsoon circulations. In coastal areas the sea breeze/land breeze cycle can define local winds; in areas that have variable terrain, mountain and valley breezes can dominate local winds.

In human civilization, the concept of wind has been explored in mythology, influenced the events of history, expanded the range of transport and warfare, and provided a power source for mechanical work, electricity and recreation. Wind powers the voyages of sailing ships across Earth's oceans.

Hot air balloons use the wind to take short trips, and powered flight uses it to increase lift and reduce fuel consumption. Areas of wind shear caused by various weather phenomena can lead to dangerous situations for aircraft. When winds become strong, trees and human-made structures are damaged or destroyed.

Winds can shape landforms, via a variety of aeolian processes such as the formation of fertile soils, such as loess, and by erosion. Dust from large deserts can be moved great distances from its source region by the prevailing winds; winds that are accelerated by rough topography and associated with dust outbreaks have been assigned regional names in various parts of the world because of their significant effects on those regions. Wind also affects the spread of wildfires. Winds can disperse seeds from various plants, enabling the survival and dispersal of those plant species, as well as flying insect populations. When combined with cold temperatures, wind has a negative impact on livestock. Wind affects animals' food stores, as well as their hunting and defensive strategies.

Causes

Wind is caused by differences in the atmospheric pressure. When a difference in atmospheric pressure exists, air moves from the higher to the lower pressure area, resulting in winds of various speeds. On a rotating planet, air will also be deflected by the Coriolis effect, except exactly on the equator. Globally, the two major driving factors of large-scale wind patterns (the atmospheric circulation) are the differential heating between the equator and the poles (difference in absorption of solar energy leading to buoyancy forces) and the rotation of the planet. Outside the tropics and aloft from frictional effects of the surface, the large-scale winds tend to approach geostrophic balance. Near the Earth's surface, friction causes the wind to be slower than it would be otherwise. Surface friction also causes winds to blow more inward into low-pressure areas.

Surface analysis of the Great Blizzard of 1888. Areas with greater isobaric packing indicate higher winds

Winds defined by an equilibrium of physical forces are used in the decomposition and analysis of wind profiles. They are useful for simplifying the atmospheric equations of motion and for making qualitative arguments about the horizontal and vertical distribution of winds. The geostrophic wind component is the result of the balance between Coriolis force and pressure gradient force. It flows parallel to isobars and approximates the flow above the atmospheric boundary layer in the midlatitudes. The thermal wind is the *difference* in the geostrophic wind between two levels in the atmosphere. It exists only in an atmosphere with horizontal temperature gradients. The ageostrophic wind component is the difference between actual and geostrophic wind, which is responsible for air "filling up" cyclones over time. The gradient wind is similar to the geostrophic wind but also includes centrifugal force (or centripetal acceleration).

Measurement

Cup-type anemometer with vertical axis, a sensor on a remote meteorological station

Wind direction is usually expressed in terms of the direction from which it originates. For example, a *northerly* wind blows from the north to the south. Weather vanes pivot to indicate the direction of the wind. At airports, windsocks indicate wind direction, and can also be used to estimate wind speed by the angle of hang. Wind speed is measured by anemometers, most commonly using rotating cups or propellers. When a high measurement frequency is needed (such as in research applications), wind can be measured by the propagation speed of ultrasound signals or by the effect of ventilation on the resistance of a heated wire. Another type of anemometer uses pitot tubes that take advantage of the pressure differential between an inner tube and an outer tube that is exposed to the wind to determine the dynamic pressure, which is then used to compute the wind speed.

An occluded mesocyclone tornado (Oklahoma, May 1999)

Sustained wind speeds are reported globally at a 10 meters (33 ft) height and are averaged over a 10 minute time frame. The United States reports winds over a 1 minute average for tropical cyclones, and a 2 minute average within weather observations. India typically reports winds over a 3minute average.

Knowing the wind sampling average is important, as the value of a one-minute sustained wind is typically 14% greater than a ten-minute sustained wind. A short burst of high speed wind is termed a wind gust, one technical definition of a wind gust is: The maxima that exceed the lowest wind speed measured during a ten-minute time interval by 10 knots (19 km/h) for periods of seconds. A squall is an increase of the wind speed above a certain threshold, which lasts for a minute or more.

To determine winds aloft, rawinsondes determine wind speed by GPS, radio navigation, or radar tracking of the probe. Alternatively, movement of the parent weather balloon position can be tracked from the ground visually using theodolites. Remote sensing techniques for wind include SODAR, Doppler lidars and radars, which can measure the Doppler shift of electromagnetic radiation scattered or reflected off suspended aerosols or molecules, and radiometers and radars can be used to measure the surface roughness of the ocean from space or airplanes. Ocean roughness can be used to estimate wind velocity close to the sea surface over oceans. Geostationary satellite imagery can be used to estimate the winds throughout the atmosphere based upon how far clouds move from one image to the next. Wind engineering describes the study of the effects of the wind on the built environment, including buildings, bridges and other man-made objects.

Wind Power

Wind energy is the kinetic energy of the air in motion. The kinetic energy of a packet of air of mass m with velocity v is given by $\frac{1}{2} m v^2$. To find the mass of the packet passing through an area A perpendicular its velocity (which could be the rotor area of a turbine), we multiply its volume after time t has passed with the air density ρ, which gives us $m = A v t \rho$. So, we find that the total wind energy is:

$$E = \frac{1}{2}\rho A v^3 t$$

Differentiating with respect to time to find the rate of increase of energy, we find that the total wind power is:

$$P = dE / dt = \frac{1}{2}\rho A v^3$$

Wind power is thus proportional to the third power of the wind velocity.

Theoretical Power Captured by a Wind Turbine

Total wind power could be captured only if the wind velocity is reduced to zero. In a realistic wind turbine this is impossible, as the captured air must also leave the turbine. A relation between the input and output wind velocity must be considered. Using the concept of stream tube, the maximal achievable extraction of wind power by a wind turbine is $16/27 \approx 59\%$ of the total theoretical wind power.

Practical Wind Turbine Power

Further insufficiencies, such as rotor blade friction and drag, gearbox losses, generator and converter losses, reduce the power delivered by a wind turbine. The basic relation that the turbine power is (approximately) proportional to the third power of velocity remains.

Global Climatology

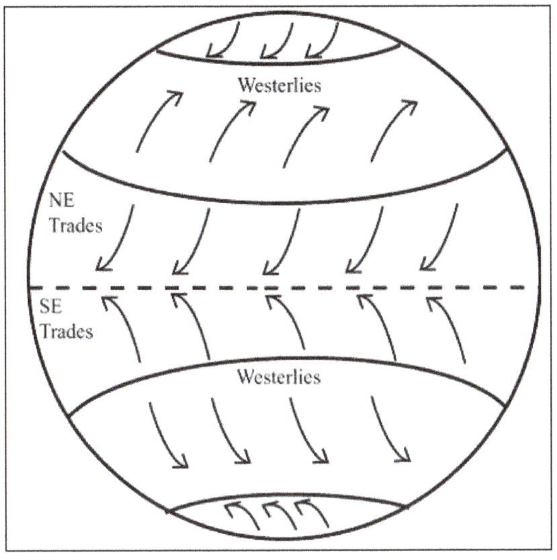

The westerlies and trade winds

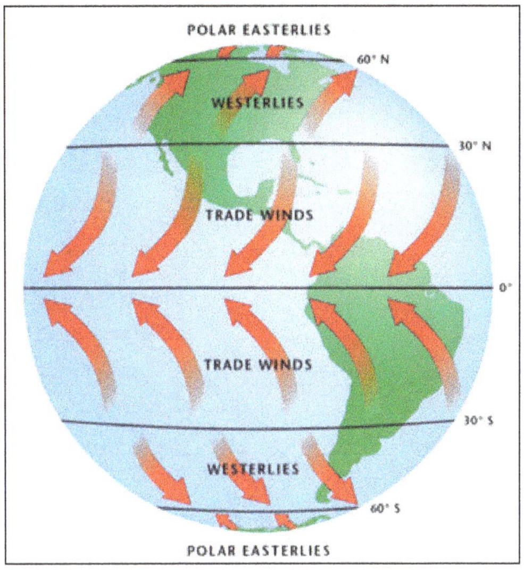

Winds are part of Earth's atmospheric circulation

Easterly winds, on average, dominate the flow pattern across the poles, westerly winds blow across the mid-latitudes of the earth, polewards of the subtropical ridge, while easterlies again dominate the tropics.

Directly under the subtropical ridge are the doldrums, or horse latitudes, where winds are lighter. Many of the Earth's deserts lie near the average latitude of the subtropical ridge, where descent reduces the relative humidity of the air mass. The strongest winds are in the mid-latitudes where cold polar air meets warm air from the tropics.

Tropics

The trade winds (also called trades) are the prevailing pattern of easterly surface winds found in the tropics towards the Earth's equator. The trade winds blow predominantly from the northeast in the Northern Hemisphere and from the southeast in the Southern Hemisphere. The trade winds act as the steering flow for tropical cyclones that form over the world's oceans. Trade winds also steer African dust westward across the Atlantic Ocean into the Caribbean, as well as portions of southeast North America.

A monsoon is a seasonal prevailing wind that lasts for several months within tropical regions. The term was first used in English in India, Bangladesh, Pakistan, and neighboring countries to refer to the big seasonal winds blowing from the Indian Ocean and Arabian Sea in the southwest bringing heavy rainfall to the area. Its poleward progression is accelerated by the development off a heat low over the Asian, African, and North American continents during May through July, and over Australia in December.

Westerlies and their Impact

The Westerlies or the Prevailing Westerlies are the prevailing winds in the middle latitudes between 35 and 65 degrees latitude. These prevailing winds blow from the west to the east, and steer extratropical cyclones in this general manner. The winds are predominantly from the southwest in the Northern Hemisphere and from the northwest in the Southern Hemisphere. They are strongest in the winter when the pressure is lower over the poles, and weakest during the summer and when pressures are higher over the poles.

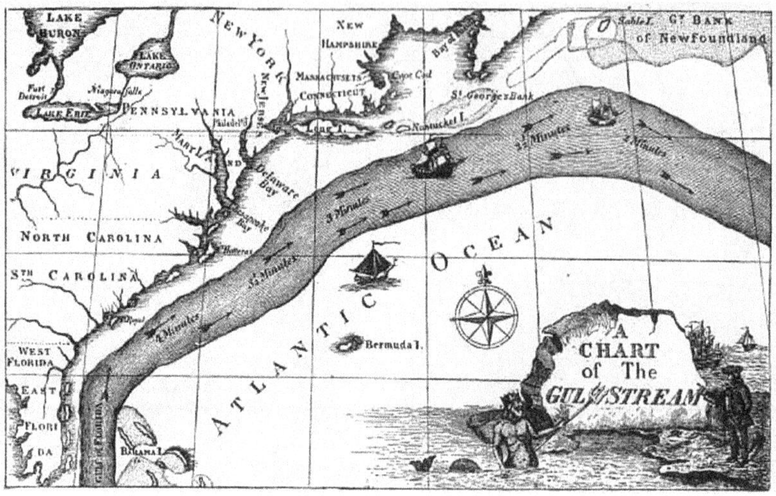

Benjamin Franklin's map of the Gulf Stream

Together with the trade winds, the westerlies enabled a round-trip trade route for sailing ships crossing the Atlantic and Pacific Oceans, as the westerlies lead to the development of strong ocean currents on the western sides of oceans in both hemispheres through the process of western intensification. These western ocean currents transport warm, sub tropical water polewards toward the polar regions. The westerlies can be particularly strong, especially in the southern hemisphere, where there is less land in the middle latitudes to cause the flow pattern to amplify, which slows the winds down. The strongest westerly winds in the middle latitudes are within a band known as the Roaring Forties, between 40 and 50 degrees latitude south of the equator. The Westerlies play an important role in carrying the warm, equatorial waters and winds to the western coasts of continents, especially in the southern hemisphere because of its vast oceanic expanse.

Polar Easterlies

The polar easterlies, also known as Polar Hadley cells, are dry, cold prevailing winds that blow from the high-pressure areas of the polar highs at the north and south poles towards the low-pressure areas within the Westerlies at high latitudes. Unlike the Westerlies, these prevailing winds blow from the east to the west, and are often weak and irregular. Because of the low sun angle, cold air builds up and subsides at the pole creating surface high-pressure areas, forcing an equatorward outflow of air; that outflow is deflected westward by the Coriolis effect.

Sea and Land Breezes

In coastal regions, sea breezes and land breezes can be important factors in a location's prevailing winds. The sea is warmed by the sun more slowly because of water's greater specific heat compared to land. As the temperature of the surface of the land rises, the land heats the air above it by conduction. The warm air is less dense than the surrounding environment and so it rises. This causes a pressure gradient of about 2 millibars from the ocean to the land. The cooler air above the sea, now with higher sea level pressure, flows inland into the lower pressure, creating a cooler breeze near the coast. When large-scale winds are calm, the strength of the sea breeze is directly proportional to the temperature difference between the land mass and the sea. If an offshore wind of 8 knots (15 km/h) exists, the sea breeze is not likely to develop.

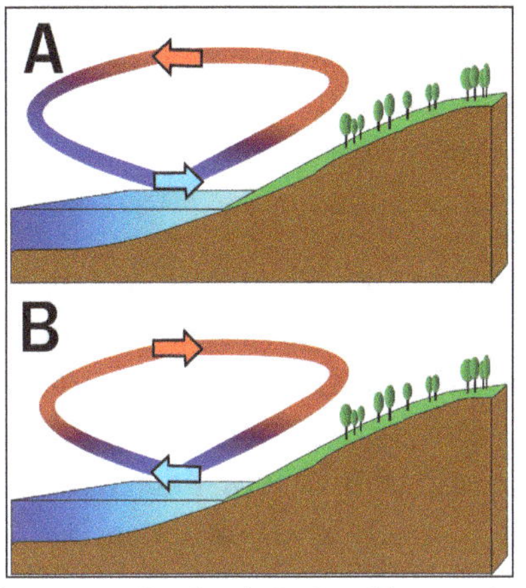

A: Sea breeze (occurs at daytime), B: Land breeze (occurs at night)

At night, the land cools off more quickly than the ocean because of differences in their specific heat values. This temperature change causes the daytime sea breeze to dissipate. When the temperature onshore cools below the temperature offshore, the pressure over the water will be lower than that of the land, establishing a land breeze, as long as an onshore wind is not strong enough to oppose it.

Near Mountains

Over elevated surfaces, heating of the ground exceeds the heating of the surrounding air at the same altitude above sea level, creating an associated thermal low over the terrain and enhancing any thermal lows that would have otherwise existed, and changing the wind circulation of the region. In areas where there is rugged topography that significantly interrupts the environmental wind flow, the wind circulation between mountains and valleys is the most important contributor to the prevailing winds. Hills and valleys substantially distort the airflow by increasing friction between the atmosphere and landmass by acting as a physical block to the flow, deflecting the wind parallel to the range just upstream of the topography, which is known as a barrier jet. This barrier jet can increase the low level wind by 45%. Wind direction also changes because of the contour of the land.

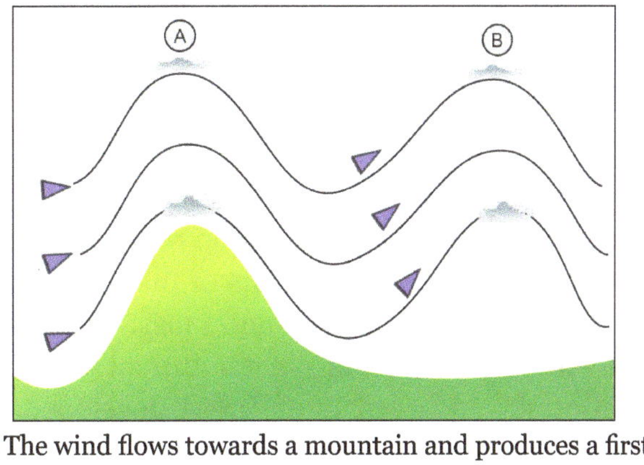

Mountain wave schematic. The wind flows towards a mountain and produces a first oscillation (A). A second wave occurs further away and higher. The lenticular clouds form at the peak of the waves (B)

If there is a pass in the mountain range, winds will rush through the pass with considerable speed because of the Bernoulli principle that describes an inverse relationship between speed and pressure. The airflow can remain turbulent and erratic for some distance downwind into the flatter countryside. These conditions are dangerous to ascending and descending airplanes. Cool winds accelerating through mountain gaps have been given regional names. In Central America, examples include the Papagayo wind, the Panama wind, and the Tehuano wind. In Europe, similar winds are known as the Bora, Tramontane, and Mistral. When these winds blow over open waters, they increase mixing of the upper layers of the ocean that elevates cool, nutrient rich waters to the surface, which leads to increased marine life.

In mountainous areas, local distortion of the airflow becomes severe. Jagged terrain combines to produce unpredictable flow patterns and turbulence, such as rotors, which can be topped by lenticular clouds. Strong updrafts, downdrafts and eddies develop as the air flows over hills and down valleys. Orographic precipitation occurs on the windward side of mountains and is caused by the rising air motion of a large-scale flow of moist air across the mountain ridge, also known as upslope flow, resulting in adiabatic cooling and condensation. In mountainous parts of the world subjected to relatively consistent winds (for example, the trade winds), a more moist climate usually prevails on the windward side of a mountain than on the leeward or downwind side. Moisture is removed by orographic lift, leaving drier air on the descending and generally warming, leeward side where a rain shadow is observed. Winds that flow over mountains down into lower elevations are known as downslope winds. These winds are warm and dry. In Europe downwind of the Alps, they are known as foehn. In Poland, an example is the halny wiatr. In Argentina, the local name for downsloped winds is zonda. In Java, the local name for such winds is koembang. In New Zealand, they are known as the Nor'west arch, and are accompanied by the cloud formation they are named after that has inspired artwork over the years. In the Great Plains of the United States, these winds are known as a chinook. Downslope winds also occur in the foothills of the Appalachian mountains of the United States, and they can be as strong as other downslope winds and unusual compared to other foehn winds in that the relative humidity typically changes little due to the increased moisture in the source air mass. In California, downslope winds are funneled through mountain passes, which intensify their effect, and examples include the Santa Ana and sundowner winds. Wind speeds during downslope wind effect can exceed 160 kilometers per hour (99 mph).

Average Wind Speeds

As described, prevailing and local winds are not spread evenly across the earth, which means that wind speeds also differ by region. In addition, the wind speed also increases with the altitude.

Wind Power Density

Nowadays, a yardstick used to determine the best locations for wind energy development is referred to as wind power density (WPD). It is a calculation relating to the effective force of the wind at a particular location, frequently expressed in terms of the elevation above ground level over a period of time. It takes into account wind velocity and mass. Color coded maps are prepared for a particular area are described as, for example, "mean annual power density at 50 meters". The results of the above calculation are included in an index developed by the National Renewable Energy Lab and referred to as "NREL CLASS". The larger the WPD calculation, the higher it is rated by class. At the end of 2008, worldwide nameplate capacity of wind-powered generators was 120.8 gigawatts. Although wind produced only about 1.5% of worldwide electricity use in 2009, it is growing rapidly, having doubled in the three years between 2005 and 2008. In several countries it has achieved relatively high levels of penetration, accounting for approximately 19% of electricity production in Denmark, 10% in Spain and Portugal, and 7% in Germany and the Republic of Ireland in 2008. One study indicates that an entirely renewable energy supply based on 70% wind is attainable at today's power prices by linking wind farms with an HVDC supergrid.Wind power has expanded quickly, its share of worldwide electricity usage at the end of 2014 was 3.1%. In 2011 wind energy was also used to power the longest journey in a wind powered car which travelled a distance of 5,000 km (3,100 miles) from Perth to Melbourne in Australia.

Shear

Wind shear, sometimes referred to as wind gradient, is a difference in wind speed and direction over a relatively short distance in the Earth's atmosphere. Wind shear can be broken down into vertical and horizontal components, with horizontal wind shear seen across weather fronts and near the coast, and vertical shear typically near the surface, though also at higher levels in the atmosphere near upper level jets and frontal zones aloft.

Wind shear itself is a microscale meteorological phenomenon occurring over a very small distance, but it can be associated with mesoscale or synoptic scale weather features such as squall lines and cold fronts. It is commonly observed near microbursts and downbursts caused by thunderstorms, weather fronts, areas of locally higher low level winds referred to as low level jets, near mountains, radiation inversions that occur because of clear skies and calm winds, buildings, wind turbines, and sailboats. Wind shear has a significant effect on the control of aircraft during take-off and landing, and was a significant cause of aircraft accidents involving large loss of life within the United States.

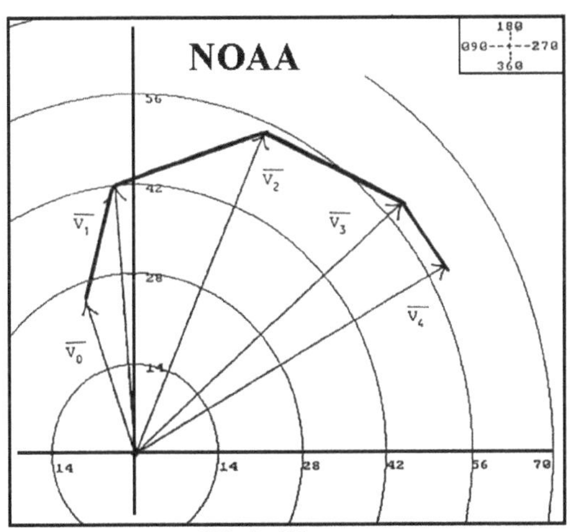

Winds / Vents

Height / Hauteur (km)	Dir	Speed / Vitesse
Sfc	161°	25
1	175°	42
2	204°	54
3	229°	59
4	241°	59

Hodograph plot of wind vectors at various heights in the troposphere, which is used to diagnose vertical wind shear

Sound movement through the atmosphere is affected by wind shear, which can bend the wave front, causing sounds to be heard where they normally would not, or vice versa. Strong vertical wind shear within the troposphere also inhibits tropical cyclone development, but helps to organize individual thunderstorms into living longer life cycles that can then produce severe weather. The thermal wind concept explains how differences in wind speed with height are dependent on horizontal temperature differences, and explains the existence of the jet stream.

Usage

Transportation

There are many different forms of sailing ships, but they all have certain basic things in common. Except for rotor ships using the Magnus effect, every sailing ship has a hull, rigging and at least one mast to hold up the sails that use the wind to power the ship. Ocean journeys by sailing ship can take many months, and a common hazard is becoming becalmed because of lack of wind, or being blown off course by severe storms or winds that do not allow progress in the desired direction. A severe storm could lead to shipwreck, and the loss of all hands. Sailing ships can only carry a certain quantity of supplies in their hold, so they have to plan long voyages carefully to include appropriate provisions, including fresh water.

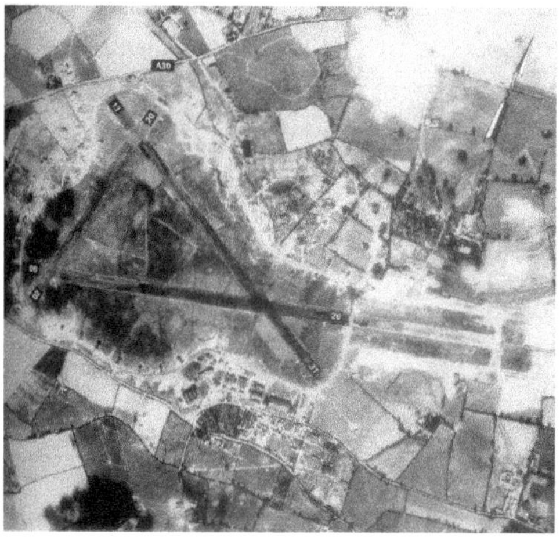

RAF Exeter airfield on 20 May 1944, showing the layout of the runways that allow aircraft to take off and land into the wind

For aerodynamic aircraft which operate relative to the air, winds affect groundspeed, and in the case of lighter-than-air vehicles, wind may play a significant or solitary role in their movement and ground track. The velocity of surface wind is generally the primary factor governing the direction of flight operations at an airport, and airfield runways are aligned to account for the common wind direction(s) of the local area. While taking off with a tailwind may be necessary under certain circumstances, a headwind is generally desirable. A tailwind increases takeoff distance required and decreases the climb gradient.

Power Source

Historically, the ancient Sinhalese of Anuradhapura and in other cities around Sri Lanka used the monsoon winds to power furnaces as early as 300 BCE. The furnaces were constructed on the path of the monsoon winds to exploit the wind power, to bring the temperatures inside up to 1,200 °C (2,190 °F). A rudimentary windmill was used to power an organ in the first century CE. The first practical windmills were later built in Sistan, Afghanistan, from the 7th century CE. These were vertical-axle windmills, which had long vertical driveshafts with rectangle shaped blades. Made of six to twelve sails covered in reed matting or cloth material, these windmills were used to grind corn and draw up water, and were used in the gristmilling and sugarcane industries. Horizontal-axle windmills were later used extensively in Northwestern Europe to grind flour beginning in the 1180s, and many Dutch windmills still exist. High altitude wind power is the focus of over 30 companies worldwide using tethered technology rather than ground-hugging compressive-towers. Oil is being saved by using wind for powering cargo ships by use of the mechanical energy converted from the wind's kinetic energy using very large kites.

This wind turbine generates electricity from wind power

Recreation

Otto Lilienthal in flight

Wind figures prominently in several popular sports, including recreational hang gliding, hot air ballooning, kite flying, snowkiting, kite landboarding, kite surfing, paragliding, sailing, and windsurfing. In gliding, wind gradients just above the surface affect the takeoff and landing phases of flight of a glider. Wind gradient can have a noticeable effect on ground launches, also known as winch launches or wire launches. If the wind gradient is significant or sudden, or both, and the pilot maintains the same pitch attitude, the indicated airspeed will increase, possibly exceeding the maximum ground launch tow speed. The pilot must adjust the airspeed to deal with the effect of the gradient. When landing, wind shear is also a hazard, particularly when the winds are strong. As the glider descends through the wind gradient on final approach to landing, airspeed decreases while sink rate increases, and there is insufficient time to accelerate prior to ground contact. The pilot must anticipate the wind gradient and use a higher approach speed to compensate for it.

Role in the Natural World

In arid climates, the main source of erosion is wind. The general wind circulation moves small particulates such as dust across wide oceans thousands of kilometers downwind of their point of origin, which is known as deflation. Westerly winds in the mid-latitudes of the planet drive the movement of ocean currents from west to east across the world's oceans. Wind has a very important role in aiding plants and other immobile organisms in dispersal of seeds, spores, pollen, etc. Although wind is not the primary form of seed dispersal in plants, it provides dispersal for a large percentage of the biomass of land plants.

Erosion

A rock formation in the Altiplano, Bolivia, sculpted by wind erosion

Erosion can be the result of material movement by the wind. There are two main effects. First, wind causes small particles to be lifted and therefore moved to another region. This is called deflation. Second, these suspended particles may impact on solid objects causing erosion by abrasion (ecological succession). Wind erosion generally occurs in areas with little or no vegetation, often in areas where there is insufficient rainfall to support vegetation. An example is the formation of sand dunes, on a beach or in a desert. Loess is a homogeneous, typically nonstratified, porous, friable, slightly coherent, often calcareous, fine-grained, silty, pale yellow or buff, windblown (Aeolian) sediment. It generally occurs as a widespread blanket deposit that covers areas of hundreds of square kilometers and tens of meters thick. Loess often stands in either steep or vertical faces. Loess tends to develop into highly rich soils. Under appropriate climatic conditions, areas with loess are among the most agriculturally productive in the world. Loess deposits are geologically unstable by nature, and will erode very readily. Therefore, windbreaks (such as big trees and bushes) are often planted by farmers to reduce the wind erosion of loess.

Desert Dust Migration

During mid-summer (July in the northern hemisphere), the westward-moving trade winds south of the northward-moving subtropical ridge expand northwestward from the Caribbean into southeastern North America. When dust from the Sahara moving around the southern periphery of the ridge within the belt of trade winds moves over land, rainfall is suppressed and the sky changes from a blue to a white appearance, which leads to an increase in red sunsets. Its presence negatively impacts air quality by adding to the count of airborne particulates. Over 50% of the African dust that reaches the United States affects Florida. Since 1970, dust outbreaks have worsened because of periods of drought in Africa. There is a large variability in the dust transport to the Caribbean and Florida from year to year. Dust events have been linked to a decline in the health of coral reefs across the Caribbean and Florida, primarily since the 1970s. Similar dust plumes originate in the Gobi Desert, which combined with pollutants, spread large distances downwind, or eastward, into North America.

There are local names for winds associated with sand and dust storms. The Calima carries dust on southeast winds into the Canary islands. The Harmattan carries dust during the winter into the Gulf of Guinea. The Sirocco brings dust from north Africa into southern Europe because of the movement of extratropical cyclones through the Mediterranean. Spring storm systems moving across the eastern Mediterranean Sea cause dust to carry across Egypt and the Arabian peninsula, which are locally known as Khamsin. The Shamal is caused by cold fronts lifting dust into the atmosphere for days at a time across the Persian Gulf states.

Effect on Plants

Wind dispersal of seeds, or anemochory, is one of the more primitive means of dispersal. Wind dispersal can take on one of two primary forms: seeds can float on the breeze or alternatively, they can flutter to the ground. The classic examples of these dispersal mechanisms include dandelions (*Taraxacum* spp., Asteraceae), which have a feathery pappus attached to their seeds and can be dispersed long distances, and maples (*Acer (genus)* spp., Sapindaceae), which have winged seeds and flutter to the ground. An important constraint on wind dispersal is the need for abundant seed production to maximize the likelihood of a seed landing in a site suitable for germination. There are also strong evolutionary constraints on this dispersal mechanism. For instance, species in the Asteraceae on islands tended to have reduced dispersal capabilities (i.e., larger seed mass and smaller pappus) relative to the same species on the mainland. Reliance upon wind dispersal is common among many weedy or ruderal species. Unusual mechanisms of wind dispersal include tumbleweeds. A related process to anemochory is anemophily, which is the process where pollen is distributed by wind. Large families of plants are pollinated in this manner, which is favored when individuals of the dominant plant species are spaced closely together.

Wind also limits tree growth. On coasts and isolated mountains, the tree line is often much lower than in corresponding altitudes inland and in larger, more complex mountain systems, because strong winds reduce tree growth. High winds scour away thin soils through erosion, as well as damage limbs and twigs. When high winds knock down or uproot trees, the process is known as windthrow. This is most likely on windward slopes of mountains, with severe cases generally occurring to tree stands that are 75 years or older. Plant varieties near the coast, such as the Sitka spruce and sea grape, are pruned back by wind and salt spray near the coastline.

Tumbleweed blown against a fence

In the montane forest of Olympic National Park, windthrow opens the canopy and increases light intensity on the understory

Wind can also cause plants damage through sand abrasion. Strong winds will pick up loose sand and topsoil and hurl it through the air at speeds ranging from 25 miles per hour (40 km/h) to 40 miles per hour (64 km/h). Such windblown sand causes extensive damage to plant seedlings because it ruptures plant cells, making them vulnerable to evaporation and drought. Using a mechanical sandblaster in a laboratory setting, scientists affiliated with the Agricultural Research Service studied the effects of windblown sand abrasion on cotton seedlings. The study showed that the seedlings responded to the damage created by the windblown sand abrasion by shifting energy from stem and root growth to the growth and repair of the damaged stems. After a period of four weeks the growth of the seedling once again became uniform throughout the plant, as it was before the windblown sand abrasion occurred.

Effect on Animals

Cattle and sheep are prone to wind chill caused by a combination of wind and cold temperatures, when winds exceed 40 kilometers per hour (25 mph), rendering their hair and wool coverings ineffective. Although penguins use both a layer of fat and feathers to help guard against coldness in both water and air, their flippers and feet are less immune to the cold. In the coldest climates such as Antarctica, emperor penguins use huddling behavior to survive the wind and cold, continuously alternating the members on the outside of the assembled group, which reduces heat loss by 50%. Flying insects, a subset of arthropods, are swept along by the prevailing winds, while birds follow their own course taking advantage of wind conditions, in order to either fly or glide. As such, fine line patterns within weather radar imagery, associated with converging winds, are dominated by insect returns. Bird migration, which tends to occur overnight within the lowest 7,000 feet (2,100 m) of the Earth's atmosphere, contaminates wind profiles gathered by weather radar, particularly the WSR-88D, by increasing the environmental wind returns by 15 knots (28 km/h) to 30 knots (56 km/h).

Pikas use a wall of pebbles to store dry plants and grasses for the winter in order to protect the food from being blown away. Cockroaches use slight winds that precede the attacks of potential predators, such as toads, to survive their encounters. Their cerci are very sensitive to the wind, and help them survive half of their attacks. Elk have a keen sense of smell that can detect potential upwind predators at a distance of 0.5 miles (800 m). Increases in wind above 15 kilometers per hour (9.3 mph) signals glaucous gulls to increase their foraging and aerial attacks on thick-billed murres.

Sound Generation

Wind causes the generation of sound. The movement of air causes movements of parts of natural objects, such as leaves or grass. These objects will produce sound if they touch each other. Even a soft wind will cause a low level of environmental noise. If the wind is blowing harder, it may produce howling sounds of varying frequencies. This may be caused by the wind blowing over cavities, or by vortices created in the air downstream of an object. Especially on high buildings, many structural parts may be a cause of annoying noise at certain wind conditions. Examples of these parts are balconies, ventilation openings, roof openings or cables.

Related Damage

High winds are known to cause damage, depending upon the magnitude of their velocity and pressure differential. Wind pressures are positive on the windward side of a structure and negative on the leeward side. Infrequent wind gusts can cause poorly designed suspension bridges to sway. When wind gusts are at a similar frequency to the swaying of the bridge, the bridge can be destroyed more easily, such as what occurred with the Tacoma Narrows Bridge in 1940. Wind speeds as low as 23 knots (43 km/h) can lead to power outages due to tree branches disrupting the flow of energy through power lines. While no species of tree is guaranteed to stand up to hurricane-force winds, those with shallow roots are more prone to uproot, and brittle trees such as eucalyptus, sea hibiscus, and avocado are more prone to damage. Hurricane-force winds cause substantial damage to mobile homes, and begin to structurally damage homes with foundations. Winds of this strength due to downsloped winds off terrain have been known to shatter windows and sandblast paint from cars. Once winds exceed 135 knots (250 km/h), homes completely collapse, and significant damage is done to larger buildings. Total destruction to man-made structures occurs when winds reach 175 knots (324 km/h). The Saffir–Simpson scale and Enhanced Fujita scale were designed to help estimate wind speed from the damage caused by high winds related to tropical cyclones and tornadoes, and vice versa.

Australia's Barrow Island holds the record for the strongest wind gust, reaching 408 km/h (253 mph) during tropical cyclone Olivia on 10 April 1996, surpassing the previous record of 372 km/h (231 mph) set on Mount Washington (New Hampshire) on the afternoon of 12 April 1934. The most powerful gusts of wind on Earth were created by nuclear detonations. The blast wave is similar to a strong wind gust over the ground. The largest nuclear explosion (50–58 megatons at an altitude of about 13,000 feet (4,000 m)) generated a 20 bar blast pressure at ground zero, which is similar to a wind gust of 3,100 miles per hour (5,000 km/h).

Damage from Hurricane Andrew

Wildfire intensity increases during daytime hours. For example, burn rates of smoldering logs are up to five times greater during the day because of lower humidity, increased temperatures, and increased wind speeds. Sunlight warms the ground during the day and causes air currents to travel uphill, and downhill during the night as the land cools. Wildfires are fanned by these winds and often follow the air currents over hills and through valleys. United States wildfire operations revolve around a 24-hour *fire day* that begins at 10:00 a.m. because of the predictable increase in intensity resulting from the daytime warmth.

STORM

A storm is any disturbed state of an environment or in an astronomical body's atmosphere especially affecting its surface, and strongly implying severe weather. It may be marked by significant disruptions to normal conditions such as strong wind, tornadoes, hail, thunder and lightning (a thunderstorm), heavy precipitation (snowstorm, rainstorm), heavy freezing rain (ice storm), strong winds (tropical cyclone, windstorm), or wind transporting some substance through the atmosphere as in a dust storm, blizzard, sandstorm, etc.

A shelf cloud, associated with a heavy or severe thunderstorm, over Swedish island of Öland in the Baltic Sea in July 2005

Storms have the potential to harm lives and property via storm surge, heavy rain or snow causing flooding or road impassibility, lightning, wildfires, and vertical wind shear. Systems with significant rainfall and duration help alleviate drought in places they move through. Heavy snowfall can allow special recreational activities to take place which would not be possible otherwise, such as skiing and snowmobiling.

Desert storms are often accompanied by violent winds, and pass rapidly.
The English word comes from Proto-Germanic *sturmaz* meaning "noise, tumult"

Lightning storm, Port-la-Nouvelle

Storms are created when a center of low pressure develops with the system of high pressure surrounding it. This combination of opposing forces can create winds and result in the formation of storm clouds such as cumulonimbus. Small localized areas of low pressure can form from hot air rising off hot ground, resulting in smaller disturbances such as dust devils and whirlwinds.

Types of Storms

Classic summer storm in Sierras de Córdoba, Argentina

A tornado in Binger, Oklahoma during the 1981 outbreak

There are many varieties and names for storms:

- Blizzard – There are varying definitions for blizzards, both over time and by location. In general, a blizzard is accompanied by gale-force winds, heavy snow (accumulating at a rate of at least 5 centimeters (2 in) per hour), and very cold conditions (below approximately −10 degrees Celsius or 14 F). Lately, the temperature criterion has fallen out of the definition across the United States.
- Bomb cyclone – A rapid deepening of a mid-latitude cyclonic low-pressure area, typically occurring over the ocean, but can occur over land. The winds experienced during these storms can be as powerful as that of a typhoon or hurricane.
- Coastal Storm – Large wind waves and/or storm surge that strike the coastal zone. Their impacts include coastal erosion and coastal flooding.
- Derecho – A derecho is a widespread, long-lived, straight-line wind storm that is associated with a land-based, fast-moving group of severe thunderstorms.
- Dust devil – A small, localized updraft of rising air.
- Dust storm – A situation in which winds pick up large quantities of sand or soil, greatly reducing the visibility.
- Firestorm – Firestorms are conflagrations which attain such intensity that they create and sustain their own wind systems. It is most commonly a natural phenomenon, created during some of the largest bushfires, forest fires, and wildfires. The Peshtigo Fire is one example of a firestorm. Firestorms can also be deliberate effects of targeted explosives such as occurred as a result of the aerial bombings of Dresden. Nuclear detonations generate firestorms if high winds are not present.
- Gale – An extratropical storm with sustained winds between 34–48 knots (39–55 mph or 63–90 km/h).
- Hailstorm – A type of storm that precipitates round chunks of ice. Hailstorms usually occur during regular thunderstorms. While most of the hail that precipitates from the clouds is fairly small and virtually harmless, there are occasional occurrences of hail greater than 2 inches (5 cm) in diameter that can cause much damage and injuries.
- Hypercane – A hypothetical tropical cyclone that could potentially form over 50 °C (122 °F) water. Such a storm would produce winds of over 800 km/h (500 mph). A series of hypercanes may have formed during the asteroid or comet impact that killed the non-avian dinosaurs 66 million years ago. Such a phenomenon could also occur during a supervolcanic eruption, or extreme global warming.
- Ice storm – Ice storms are one of the most dangerous forms of winter storms. When surface temperatures are below freezing, but a thick layer of above-freezing air remains aloft, rain can fall into the freezing layer and freeze upon impact into a glaze of ice. In general, 8 millimetres (0.31 in) of accumulation is all that is required, especially in combination with breezy conditions, to start downing power lines as well as tree limbs. Ice storms also make unheated road surfaces too slick to drive upon. Ice storms can vary in time range from hours to days and can cripple small towns and large metropolitan cities alike.

- Microburst – A very powerful windstorm produced during a thunderstorm that only lasts a few minutes.
- Ocean Storm or sea storm – Storm conditions out at sea are defined as having sustained winds of 48 knots (55 mph or 90 km/h) or greater. Usually just referred to as a storm, these systems can sink vessels of all types and sizes.
- Snowstorm – A heavy fall of snow accumulating at a rate of more than 5 centimeters (2 in) per hour that lasts several hours. Snow storms, especially ones with a high liquid equivalent and breezy conditions, can down tree limbs, cut off power connections and paralyze travel over large regions.
- Squall – Sudden onset of wind increase of at least 16 knots (30 km/h) or greater sustained for at least one minute.
- Thunderstorm – A thunderstorm is a type of storm that generates both lightning and thunder. It is normally accompanied by heavy precipitation. Thunderstorms occur throughout the world, with the highest frequency in tropical rainforest regions where there are conditions of high humidity and temperature along with atmospheric instability. These storms occur when high levels of condensation form in a volume of unstable air that generates deep, rapid, upward motion in the atmosphere. The heat energy creates powerful rising air currents that swirl upwards to the tropopause. Cool descending air currents produce strong downdraughts below the storm. After the storm has spent its energy, the rising currents die away and downdraughts break up the cloud. Individual storm clouds can measure 2–10 km across.
- Tornado – A tornado is a violent, destructive whirlwind storm occurring on land. Usually its appearance is that of a dark, funnel-shaped cloud. Often tornadoes are preceded by or associated with thunderstorms and a wall cloud. They are often called the most destructive of storms, and while they form all over the planet, the interior of the United States is the most prone area, especially throughout Tornado Alley.
- Tropical cyclone – A tropical cyclone is a storm system with a closed circulation around a centre of low pressure, fueled by the heat released when moist air rises and condenses. The name underscores its origin in the tropics and their cyclonic nature. Tropical cyclones are distinguished from other cyclonic storms such as nor'easters and polar lows by the heat mechanism that fuels them, which makes them "warm core" storm systems. Tropical cyclones form in the oceans if the conditions in the area are favorable, and depending on their strength and location, there are various terms by which they are called, such as *tropical depression*, *tropical storm*, *hurricane* and *typhoon*.
- Wind storm – A storm marked by high wind with little or no precipitation. Windstorm damage often opens the door for massive amounts of water and debris to cause further damage to a structure. European windstorms and derechos are two type of windstorms. High wind is also the cause of sandstorms in dry climates.

Classification of Storms

A strict meteorological definition of a terrestrial storm is a wind measuring 10 or higher on the

Beaufort scale, meaning a wind speed of 24.5 m/s (89 km/h, 55 mph) or more; however, popular usage is not so restrictive. Storms can last anywhere from 12 to 200 hours, depending on season and geography. In North America, the east and northeast storms are noted for the most frequent repeatability and duration, especially during the cold period. Big terrestrial storms alter the oceanographic conditions that in turn may affect food abundance and distribution: strong currents, strong tides, increased siltation, change in water temperatures, overturn in the water column, etc.

Extraterrestrial Storms

Storms do not only occur on Earth; other planetary bodies with a sufficient atmosphere (gas giants in particular) also undergo stormy weather. The Great Red Spot on Jupiter provides a well-known example. Though technically an anticyclone, with greater than hurricane wind speeds, it is larger than the Earth and has persisted for at least 340 years, having first been observed by astronomer Galileo Galilei. Neptune also had its own lesser-known Great Dark Spot.

The Great Red Spot on Jupiter

In September 1994, the Hubble telescope – using Wide Field Planetary Camera 2 – imaged storms on Saturn generated by upwelling of warmer air, similar to a terrestrial thunderhead. The east-west extent of the same-year storm equalled the diameter of Earth. The storm was observed earlier in September 1990 and acquired the name Dragon Storm.

The dust storms of Mars vary in size, but can often cover the entire planet. They tend to occur when Mars comes closest to the Sun, and have been shown to increase the global temperature.

One particularly large Martian storm was exhaustively studied up close due to coincidental timing. When the first spacecraft to successfully orbit another planet, Mariner 9, arrived and successfully orbited Mars on 14 November 1971, planetary scientists were surprised to find the atmosphere was thick with a planet-wide robe of dust, the largest storm ever observed on Mars. The surface of the planet was totally obscured. Mariner 9's computer was reprogrammed from Earth to delay imaging of the surface for a couple of months until the dust settled, however, the surface-obscured images contributed much to the collection of Mars atmospheric and planetary surface science.

Two extrasolar planets are known to have storms: HD 209458 b and HD 80606 b. The former's storm was discovered on June 23, 2010 and measured at 6,200 km/h, while the latter produces winds of 17,700 kilometers (11,000 mi) per hour across the surface. The spin of the planet then creates giant swirling shock-wave storms that carry the heat aloft.

Effects on Human Society

A return stroke, cloud-to-ground lightning strike during a thunderstorm

Shipwrecks are common with the passage of strong tropical cyclones. Such shipwrecks can change the course of history, as well as influence art and literature. A hurricane led to a victory of the Spanish over the French for control of Fort Caroline, and ultimately the Atlantic coast of North America, in 1565.

A snow blockade in southern Minnesota in 1881

A sunshower storm in the Mojave desert at sunset

Strong winds from any storm type can damage or destroy vehicles, buildings, bridges, and other outside objects, turning loose debris into deadly flying projectiles. In the United States, major hurricanes comprise just 21% of all landfalling tropical cyclones, but account for 83% of all damage.

Tropical cyclones often knock out power to tens or hundreds of thousands of people, preventing vital communication and hampering rescue efforts. Tropical cyclones often destroy key bridges, overpasses, and roads, complicating efforts to transport food, clean water, and medicine to the areas that need it. Furthermore, the damage caused by tropical cyclones to buildings and dwellings can result in economic damage to a region, and to a diaspora of the population of the region.

The storm surge, or the increase in sea level due to the cyclone, is typically the worst effect from landfalling tropical cyclones, historically resulting in 90% of tropical cyclone deaths. The relatively quick surge in sea level can move miles/kilometers inland, flooding homes and cutting off escape routes. The storm surges and winds of hurricanes may be destructive to human-made structures, but they also stir up the waters of coastal estuaries, which are typically important fish breeding locales.

Cloud-to-ground lightning frequently occurs within the phenomena of thunderstorms and have numerous hazards towards landscapes and populations. One of the more significant hazards lightning can pose is the wildfires they are capable of igniting. Under a regime of low precipitation (LP) thunderstorms, where little precipitation is present, rainfall cannot prevent fires from starting when vegetation is dry as lightning produces a concentrated amount of extreme heat. Wildfires can devastate vegetation and the biodiversity of an ecosystem. Wildfires that occur close to urban environments can inflict damages upon infrastructures, buildings, crops, and provide risks to explosions, should the flames be exposed to gas pipes. Direct damage caused by lightning strikes occurs on occasion. In areas with a high frequency for cloud-to-ground lightning, like Florida, lightning causes several fatalities per year, most commonly to people working outside.

Precipitation with low potential of hydrogen levels (pH), otherwise known as acid rain, is also a frequent risk produced by lightning. Distilled water, which contains no carbon dioxide, has a neutral pH of 7. Liquids with a pH less than 7 are acidic, and those with a pH greater than 7 are bases. "Clean" or unpolluted rain has a slightly acidic pH of about 5.2, because carbon dioxide and water in the air react together to form carbonic acid, a weak acid (pH 5.6 in distilled water), but unpolluted rain also contains other chemicals. Nitric oxide present during thunderstorm phenomena, caused by the splitting of nitrogen molecules, can result in the production of acid rain, if nitric oxide forms compounds with the water molecules in precipitation, thus creating acid rain. Acid rain can damage infrastructures containing calcite or other solid chemical compounds containing carbon. In ecosystems, acid rain can dissolve plant tissues of vegetations and increase acidification process in bodies of water and in soil, resulting in deaths of marine and terrestrial organisms.

Hail damage to roofs often goes unnoticed until further structural damage is seen, such as leaks or cracks. It is hardest to recognize hail damage on shingled roofs and flat roofs, but all roofs have their own hail damage detection problems. Metal roofs are fairly resistant to hail damage, but may accumulate cosmetic damage in the form of dents and damaged coatings. Hail is also a common nuisance to drivers of automobiles, severely denting the vehicle and cracking or even shattering windshields and windows. Rarely, massive hailstones have been known to cause concussions or fatal head trauma. Hailstorms have been the cause of costly and deadly events throughout history. One of the earliest recorded incidents occurred around the 9th century in Roopkund, Uttarakhand, India. The largest hailstone in terms of diameter and weight ever recorded in the United States fell on July 23, 2010 in Vivian, South Dakota in the United States; it measured 8 inches (20 cm) in diameter and 18.62 inches (47.3 cm) in circumference, weighing in at 1.93 pounds (0.88 kg). This broke the previous record for diameter set by a hailstone 7 inches diameter and 18.75 inches circumference which fell in Aurora, Nebraska in the United States on June 22, 2003, as well as the record for weight, set by a hailstone of 1.67 pounds (0.76 kg) that fell in Coffeyville, Kansas in 1970.

Various hazards, ranging from hail to lightning can affect outside technology facilities, such as antennas, satellite dishes, and towers. As a result, companies with outside facilities have begun installing such facilities underground, in order to reduce the risk of damage from storms.

Substantial snowfall can disrupt public infrastructure and services, slowing human activity even in regions that are accustomed to such weather. Air and ground transport may be greatly inhibited or shut down entirely. Populations living in snow-prone areas have developed various ways to travel across the snow, such as skis, snowshoes, and sleds pulled by horses, dogs, or other animals and later, snowmobiles. Basic utilities such as electricity, telephone lines, and gas supply can also fail. In addition, snow can make roads much harder to travel and vehicles attempting to use them can easily become stuck.

The combined effects can lead to a "snow day" on which gatherings such as school, work, or church are officially canceled. In areas that normally have very little or no snow, a snow day may occur when there is only light accumulation or even the threat of snowfall, since those areas are unprepared to handle any amount of snow. In some areas, such as some states in the United States, schools are given a yearly quota of snow days (or "calamity days"). Once the quota is exceeded, the snow days must be made up. In other states, all snow days must be made up. For example, schools may extend the remaining school days later into the afternoon, shorten spring break, or delay the start of summer vacation.

Accumulated snow is removed to make travel easier and safer, and to decrease the long-term effect of a heavy snowfall. This process utilizes shovels and snowplows, and is often assisted by sprinkling salt or other chloride-based chemicals, which reduce the melting temperature of snow. In some areas with abundant snowfall, such as Yamagata Prefecture, Japan, people harvest snow and store it surrounded by insulation in ice houses. This allows the snow to be used through the summer for refrigeration and air conditioning, which requires far less electricity than traditional cooling methods.

Agriculture

Hail can cause serious damage, notably to automobiles, aircraft, skylights, glass-roofed structures, livestock, and most commonly, farmers' crops. Wheat, corn, soybeans, and tobacco are the most sensitive crops to hail damage. Hail is one of Canada's most expensive hazards. Snowfall can be beneficial to agriculture by serving as a thermal insulator, conserving the heat of the Earth and protecting crops from subfreezing weather. Some agricultural areas depend on an accumulation of snow during winter that will melt gradually in spring, providing water for crop growth. If it melts into water and refreezes upon sensitive crops, such as oranges, the resulting ice will protect the fruit from exposure to lower temperatures. Although tropical cyclones take an enormous toll in lives and personal property, they may be important factors in the precipitation regimes of places they affect and bring much-needed precipitation to otherwise dry regions. Hurricanes in the eastern north Pacific often supply moisture to the Southwestern United States and parts of Mexico. Japan receives over half of its rainfall from typhoons.

Hurricane Camille averted drought conditions and ended water deficits along much of its path, though it also killed 259 people and caused $9.14 billion (2005 USD) in damage.

Aviation

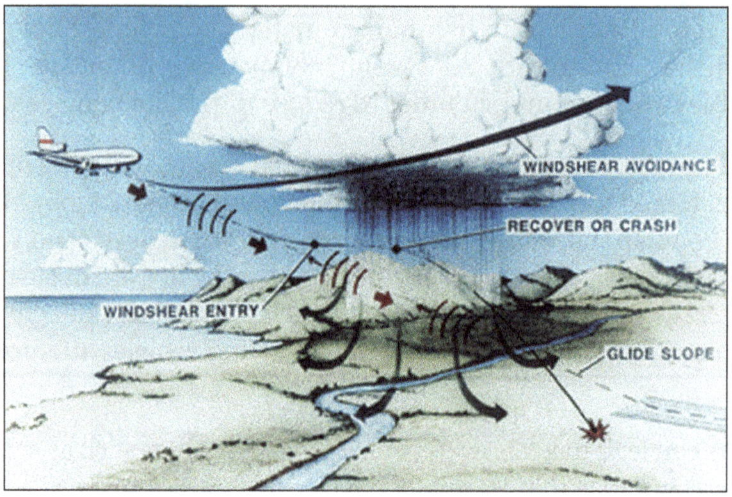

Effect of wind shear on aircraft trajectory. Merely correcting for the initial gust front can have dire consequences

Hail is one of the most significant thunderstorm hazards to aircraft. When hail stones exceed 0.5 inches (13 mm) in diameter, planes can be seriously damaged within seconds. The hailstones accumulating on the ground can also be hazardous to landing aircraft. Strong wind outflow from thunderstorms causes rapid changes in the three-dimensional wind velocity just above ground level. Initially, this outflow causes a headwind that increases airspeed, which normally causes a pilot to reduce engine power if they are unaware of the wind shear. As the aircraft passes into the region of the downdraft, the localized headwind diminishes, reducing the aircraft's airspeed and increasing its sink rate. Then, when the aircraft passes through the other side of the downdraft, the headwind becomes a tailwind, reducing lift generated by the wings, and leaving the aircraft in a low-power, low-speed descent. This can lead to an accident if the aircraft is too low to effect a recovery before ground contact. As the result of the accidents in the 1970s and 1980s, in 1988 the U.S. Federal Aviation Administration mandated that all commercial aircraft have on-board wind shear detection systems by 1993. Between 1964 and 1985, wind shear directly caused or contributed to 26 major civil transport aircraft accidents in the U.S. that led to 620 deaths and 200 injuries. Since 1995, the number of major civil aircraft accidents caused by wind shear has dropped to approximately one every ten years, due to the mandated on-board detection as well as the addition of Doppler weather radar units on the ground.

SNOW

Snow comprises individual ice crystals that grow while suspended in the atmosphere—usually within clouds—and then fall, accumulating on the ground where they undergo further changes. It consists of frozen crystalline water throughout its life cycle, starting when, under suitable conditions, the ice crystals form in the atmosphere, increase to millimeter size, precipitate and accumulate on surfaces, then metamorphose in place, and ultimately melt, slide or sublimate away. Snowstorms organize and develop by feeding on sources of atmospheric moisture and cold air. Snowflakes nucleate around particles in the atmosphere by attracting supercooled water droplets, which freeze in hexagonal-shaped crystals. Snowflakes take on a variety of shapes, basic among these are platelets, needles, columns and rime. As snow accumulates into a snowpack, it may blow into drifts. Over time, accumulated snow metamorphoses, by sintering, sublimation and freeze-thaw. Where the climate is cold enough for year-to-year accumulation, a glacier may form. Otherwise, snow typically melts seasonally, causing runoff into streams and rivers and recharging groundwater.

Major snow-prone areas include the polar regions, the northernmost half of the Northern Hemisphere and mountainous regions worldwide with sufficient moisture and cold temperatures. In the Southern Hemisphere, snow is confined primarily to mountainous areas, apart from Antarctica.

Snow affects such human activities as transportation: creating the need for keeping roadways, wings, and windows clear; agriculture: providing water to crops and safeguarding livestock; sports such as skiing, snowboarding, and snowmachine travel; and warfare. Snow affects ecosystems, as well, by providing an insulating layer during winter under which plants and animals are able to survive the cold.

Precipitation

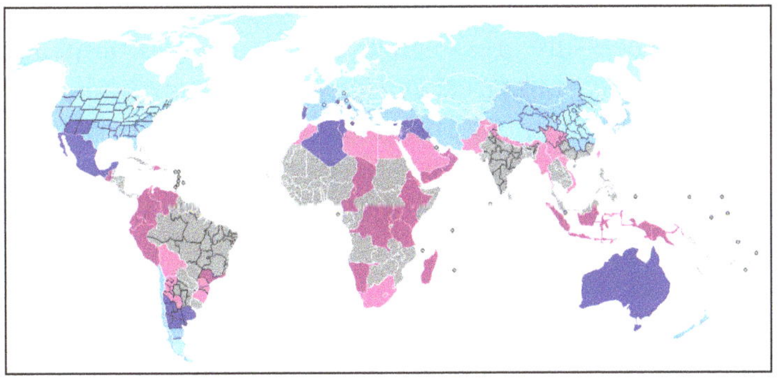

Worldwide occurrence of snowfall. Snow at reference above sea level (meters):

- Below 500: annually.
- Below 500: annually, but not in all of its territory.
- 500: above annually, below occasionally.
- Above 500: annually.
- Above 2,000: annually.
- Any elevation: none.

Snow develops in clouds that themselves are part of a larger weather system. The physics of snow crystal development in clouds results from a complex set of variables that include moisture content and temperatures. The resulting shapes of the falling and fallen crystals can be classified into a number of basic shapes and combinations, thereof. Occasionally, some plate-like, dendritic and stellar-shaped snowflakes can form under clear sky with a very cold temperature inversion present.

Cloud Formation

Snow clouds usually occur in the context of larger weather systems, the most important of which is the low pressure area, which typically incorporate warm and cold fronts as part of their circulation. Two additional and locally productive sources of snow are lake-effect (also sea-effect) storms and elevation effects, especially in mountains.

Low Pressure Areas

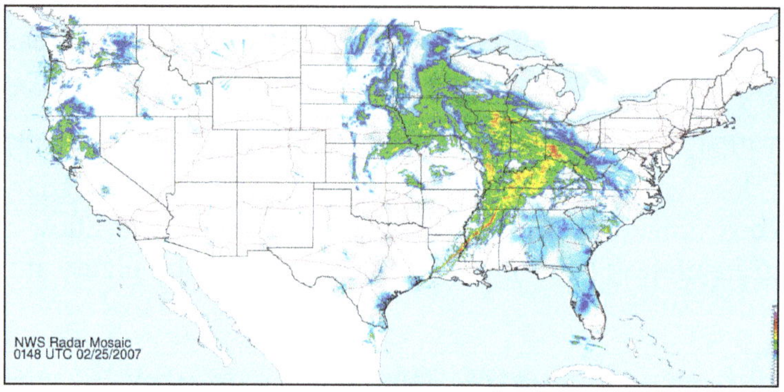

Extratropical cyclonic snowstorm, February 24, 2007

Mid-latitude cyclones are low pressure areas which are capable of producing anything from cloudiness and mild snow storms to heavy blizzards. During a hemisphere's fall, winter, and spring, the atmosphere over continents can be cold enough through the depth of the troposphere to cause snowfall. In the Northern Hemisphere, the northern side of the low pressure area produces the most snow. For the southern mid-latitudes, the side of a cyclone that produces the most snow is the southern side.

Fronts

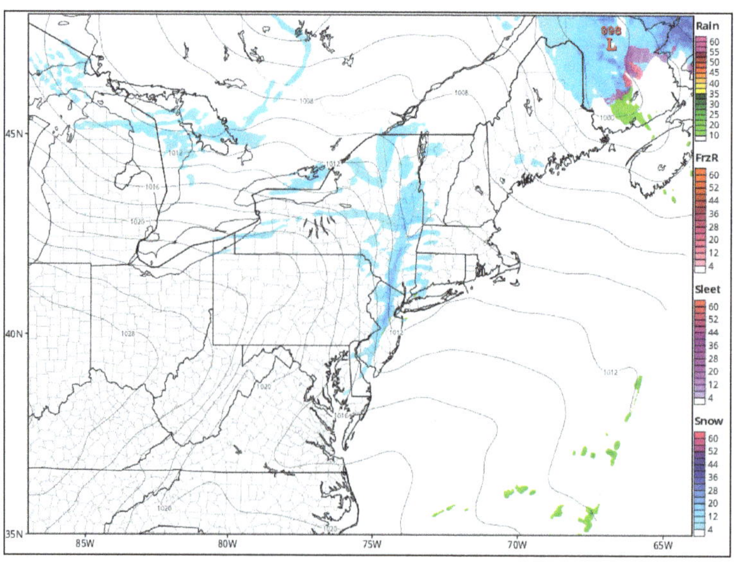

Frontal snowsquall moving toward Boston, Massachusetts.

A cold front, the leading edge of a cooler mass of air, can produce frontal snowsqualls—an intense frontal convective line (similar to a rainband), when temperature is near freezing at the surface. The strong convection that develops has enough moisture to produce whiteout conditions at places which line passes over as the wind causes intense blowing snow. This type of snowsquall generally lasts less than 30 minutes at any point along its path but the motion of the line can cover large distances. Frontal squalls may form a short distance ahead of the surface cold front or behind the cold front where there may be a deepening low pressure system or a series of trough lines which act similar to a traditional cold frontal passage. In situations where squalls develop post-frontally it is not unusual to have two or three linear squall bands pass in rapid succession only separated by 25 miles (40 kilometers) with each passing the same point in roughly 30 minutes apart. In cases where there is a large amount of vertical growth and mixing the squall may develop embedded cumulonimbus clouds resulting in lightning and thunder which is dubbed thundersnow.

A warm front can produce snow for a period, as warm, moist air overrides below-freezing air and creates precipitation at the boundary. Often, snow transitions to rain in the warm sector behind the front.

Lake and Ocean Effects

Lake-effect snow is produced during cooler atmospheric conditions when a cold air mass moves across long expanses of warmer lake water, warming the lower layer of air which picks up water vapor from the lake, rises up through the colder air above, freezes and is deposited on the leeward (downwind) shores.

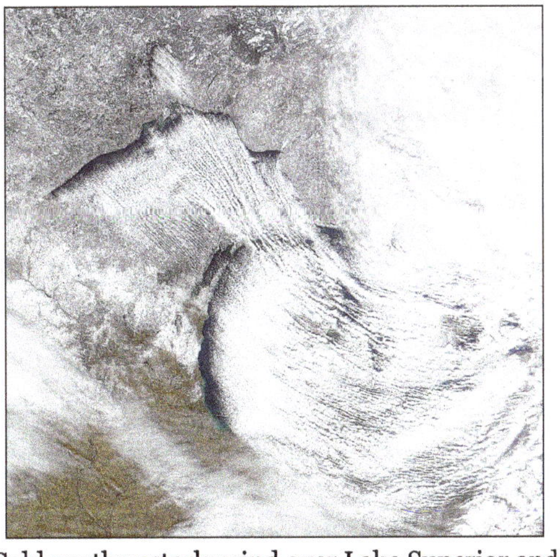

Cold northwesterly wind over Lake Superior and Lake Michigan creating lake-effect snowfall

The same effect also occurs over bodies of salt water, when it is termed *ocean-effect* or *bay-effect snow*. The effect is enhanced when the moving air mass is uplifted by the orographic influence of higher elevations on the downwind shores. This uplifting can produce narrow but very intense bands of precipitation, which deposit at a rate of many inches of snow each hour, often resulting in a large amount of total snowfall.

The areas affected by lake-effect snow are called snowbelts. These include areas east of the Great Lakes, the west coasts of northern Japan, the Kamchatka Peninsula in Russia, and areas near the Great Salt Lake, Black Sea, Caspian Sea, Baltic Sea, and parts of the northern Atlantic Ocean.

Mountain Effects

Orographic or relief snowfall is caused when masses of air pushed by wind are forced up the side of elevated land formations, such as large mountains. The lifting of air up the side of a mountain or range results in adiabatic cooling, and ultimately condensation and precipitation. Moisture is removed by orographic lift, leaving drier, warmer air on the descending, leeward side. The resulting enhanced productivity of snow fall and the decrease in temperature with elevation means that snow depth and seasonal persistence of snowpack increases with elevation in snow-prone areas.

Cloud Physics

Freshly fallen snowflakes

A snowflake consists of roughly 10^{19} water molecules, which are added to its core at different rates and in different patterns, depending on the changing temperature and humidity within the atmosphere that the snowflake falls through on its way to the ground. As a result, snowflakes vary among themselves, while following similar patterns.

Snow crystals form when tiny supercooled cloud droplets (about 10 μm in diameter) freeze. These droplets are able to remain liquid at temperatures lower than −18 °C (0 °F), because to freeze, a few molecules in the droplet need to get together by chance to form an arrangement similar to that in an ice lattice. Then the droplet freezes around this "nucleus". In warmer clouds an aerosol particle or "ice nucleus" must be present in (or in contact with) the droplet to act as a nucleus. Ice nuclei are very rare compared to that cloud condensation nuclei on which liquid droplets form. Clays, desert dust and biological particles can be nuclei. Artificial nuclei include particles of silver iodide and dry ice, and these are used to stimulate precipitation in cloud seeding.

Once a droplet has frozen, it grows in the supersaturated environment—one where air is saturated with respect to ice when the temperature is below the freezing point. The droplet then grows by diffusion of water molecules in the air (vapor) onto the ice crystal surface where they are collected. Because water droplets are so much more numerous than the ice crystals due to their sheer abundance, the crystals are able to grow to hundreds of micrometers or millimeters in size at the expense of the water droplets by the Wegener–Bergeron–Findeisen process. The corresponding depletion of water vapor causes the ice crystals to grow at the droplets' expense. These large crystals are an efficient source of precipitation, since they fall through the atmosphere due to their mass, and may collide and stick together in clusters, or aggregates. These aggregates are snowflakes, and are usually the type of ice particle that falls to the ground. Although the ice is clear, scattering of light by the crystal facets and hollows/imperfections mean that the crystals often appear white in color due to diffuse reflection of the whole spectrum of light by the small ice particles.

Classification of Snowflakes

An early classification of snowflakes by Israel Perkins Warren

Micrography of thousands of snowflakes from 1885 onward, starting with Wilson Alwyn Bentley, revealed the wide diversity of snowflakes within a classifiable set of patterns. Closely matching snow crystals have been observed.

Ukichiro Nakaya developed a crystal morphology diagram, relating crystal shapes to the temperature and moisture conditions under which they formed, which is summarized in the following table.

Crystal structure morphology as a function of temperature and water saturation					
Temperature range		Saturation range		Types of snow crystal	
°C	°F	g/m³	oz/cu yd	below saturation	above saturation
0 to −3.5	32 to 26	0.0 to 0.5	0.000 to 0.013	Solid plates	Thin plates Dendrites
−3.5 to −10	26 to 14	0.5 to 1.2	0.013 to 0.032	Solid prisms Hollow prisms	Hollow prisms Needles
−10 to −22	14 to −8	1.2 to 1.4	0.032 to 0.038	Thin plates Solid plates	Sectored plates Dendrites
−22 to −40	−8 to −40	1.2 to 0.1	0.0324 to 0.0027	Thin plates Solid plates	Columns Prisms

Nakaya discovered that the shape is also a function of whether the prevalent moisture is above or below saturation. Forms below the saturation line trend more towards solid and compact. Crystals formed in supersaturated air trend more towards lacy, delicate and ornate. Many more complex growth patterns also form such as side-planes, bullet-rosettes and also planar types depending on the conditions and ice nuclei. If a crystal has started forming in a column growth regime, at around −5 °C (23 °F), and then falls into the warmer plate-like regime, then plate or dendritic crystals sprout at the end of the column, producing so called "capped columns".

Magono and Lee devised a classification of freshly formed snow crystals that includes 80 distinct shapes. They documented each with micrographs.

Accumulation

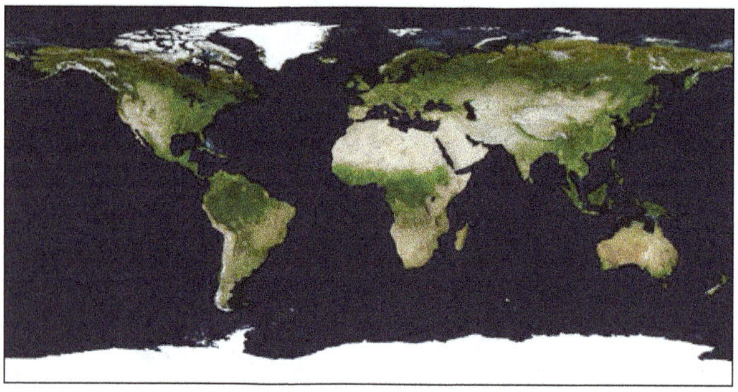

An animation of seasonal snow changes, based on satellite imagery

Snow accumulates from a series of snow events, punctuated by freezing and thawing, over areas that are cold enough to retain snow seasonally or perennially. Major snow-prone areas include the Arctic and Antarctic, the Northern Hemisphere, and alpine regions. The liquid equivalent of snowfall may be evaluated using a snow gauge or with a standard rain gauge, adjusted for winter by removal of a funnel and inner cylinder. Both types of gauges melt the accumulated snow and report the amount of water collected. At some automatic weather stations an ultrasonic snow depth sensor may be used to augment the precipitation gauge.

Snow Events

Snow flurry, snow storm and blizzard describe snow events of progressively greater duration and intensity. A blizzard is a weather condition involving snow and has varying definitions in different parts of the world. In the United States, a blizzard occurs when two conditions are met for a period of three hours or more: A sustained wind or frequent gusts to 35 miles per hour (56 km/h), and sufficient snow in the air to reduce visibility to less than 0.4 kilometers (0.25 mi). In Canada and the United Kingdom, the criteria are similar. While heavy snowfall often occurs during blizzard conditions, falling snow is not a requirement, as blowing snow can create a ground blizzard.

Snowstorm intensity may be categorized by visibility and depth of accumulation. Snowfall's intensity is determined by visibility, as follows:

- *Light*: visibility greater than 1 kilometer (0.6 mi).
- *Moderate*: visibility restrictions between 0.5 and 1 kilometer (0.3 and 0.6 mi).
- *Heavy*: visibility is less than 0.5 kilometers (0.3 mi).

The *International Classification for Seasonal Snow on the Ground* defines "height of new snow" as the depth of freshly fallen snow, in centimeters as measured with a ruler, that accumulated on a snowboard during an observation period of 24 hours, or other observation interval. After the measurement, the snow is cleared from the board and the board is placed flush with the snow surface to provide an accurate measurement at the end of the next interval. Melting, compacting, blowing and drifting contribute to the difficulty of measuring snowfall.

Distribution

Glaciers with their permanent snowpacks cover about 10% of the earth's surface, while seasonal snow covers about nine percent, mostly in the Northern Hemisphere, where seasonal snow covers about 40 million square kilometres (15×10^6 sq mi), according to a 1987 estimate. A 2007 estimate of snow cover over the Northern Hemisphere suggested that, on average, snow cover ranges from a minimum extent of 2 million square kilometres (0.77×10^6 sq mi) each August to a maximum extent of 45 million square kilometres (17×10^6 sq mi) each January or nearly half of the land surface in that hemisphere. A study of Northern Hemisphere snow cover extent for the period 1972–2006 suggests a reduction of 0.5 million square kilometres (0.19×10^6 sq mi) over the 35-year period.

Records

The following are world records regarding snowfall and snowflakes:

- *Highest seasonal total snowfall* – The world record for the highest seasonal total snowfall was measured in the United States at Mt. Baker Ski Area, outside of the city of Bellingham, Washington during the 1998–1999 season. Mount Baker received 2,896 cm (95.01 ft) of snow, thus surpassing the previous record holder, Mount Rainier, Washington, which during the 1971–1972 season received 2,850 cm (93.5 ft) of snow.
- *Highest seasonal average annual snowfall* – The world record for the highest average annual snowfall is 1,764 cm (57.87 ft), measured in Sukayu Onsen, Japan for the period of 1981–2010.
- *Largest snowflake* – According to *Guinness World Records,* the world's largest snowflake fell in January 1887 outside present-day Miles City, Montana. It measured 38 cm (15 in) in diameter.

Metamorphosis

Fresh snow beginning to metamorphose: The surface shows wind packing and sastrugi. In the foreground are hoar frost crystals, formed by refrozen water vapor emerging to the cold surface

After deposition, snow progresses on one of two paths that determine its fate, either *ablation* (mostly by melting) or transitioning from firn (multi-year snow) into *glacier ice*. During this transition, snow "is a highly porous, sintered material made up of a continuous ice structure and a continuously connected pore space, forming together the snow microstructure". Almost always near its melting temperature, a snowpack is continually transforming these properties in a process, known as *metamorphism*, wherein all three phases of water may coexist, including liquid water partially filling the pore space. Starting as a powdery deposition, snow becomes more granular when it begins to compact under its own weight, be blown by the wind, sinter particles together and commence the cycle of melting and refreezing. Water vapor plays a role as it deposits ice crystals, known as hoar frost, during cold, still conditions.

Seasonal Snowpack

Over the course of time, a snowpack may settle under its own weight until its density is approximately 30% of water. Increases in density above this initial compression occur primarily by melting and refreezing, caused by temperatures above freezing or by direct solar radiation. In colder climates, snow lies on the ground all winter. By late spring, snow densities typically reach a maximum of 50% of water. Snow that persists into summer evolves into névé, granular snow, which has been partially melted, refrozen and compacted. Névé has a minimum density of 500 kilograms per cubic metre (31 lb/cu ft), which is roughly half of the density of liquid water.

Firn

Firn is snow that has persisted for multiple years and has been recrystallized into a substance denser than névé, yet less dense and hard than glacial ice. Firn resembles caked sugar and is very resistant to shovelling. Its density generally ranges from 550 kilograms per cubic metre (34 lb/cu ft) to 830 kilograms per cubic metre (52 lb/cu ft), and it can often be found underneath the snow that accumulates at the head of a glacier. The minimum altitude that firn accumulates on a glacier is called the *firn limit*, *firn line* or *snowline*.

Firn—metamorphosed multi-year snow

Movement

There are four main mechanisms for movement of deposited snow: *drifting* of unsintered snow, *avalanches* of accumulated snow on steep slopes, *snowmelt* during thaw conditions, and the *movement of glaciers* after snow has persisted for multiple years and metamorphosed into glacier ice.

Drifting

Snow drifts forming around downwind obstructions

When powdery, snow drifts with the wind from the location where it originally fell, forming deposits with a depth of several meters in isolated locations. After attaching to hillsides, blown snow can evolve into a snow slab, which is an avalanche hazard on steep slopes.

Avalanche

An avalanche (also called a snowslide or snowslip) is a rapid flow of snow down a sloping surface. Avalanches are typically triggered in a starting zone from a mechanical failure in the snowpack (slab avalanche) when the forces on the snow exceed its strength but sometimes only with gradually widening (loose snow avalanche). After initiation, avalanches usually accelerate rapidly and grow in mass and volume as they entrain more snow. If the avalanche moves fast enough some of the snow may mix with the air forming a powder snow avalanche, which is a type of gravity current. They occur in three major mechanisms:

- *Slab avalanches* occur in snow that has been deposited, or redeposited by wind. They have the characteristic appearance of a block (slab) of snow cut out from its surroundings by fractures. These account for most back-country fatalities.
- *Powder snow avalanches* result from a deposition of fresh dry powder and generate a powder cloud, which overlies a dense avalanche. They can exceed speeds of 300 kilometres per hour (190 mph), and masses of 10,000,000 tonnes (9,800,000 long tons; 11,000,000 short tons); their flows can travel long distances along flat valley bottoms and even uphill for short distances.

A powder snow avalanche

- *Wet snow avalanches* are a low-velocity suspension of snow and water, with the flow confined to the surface of the pathway. The low speed of travel is due to the friction between the sliding surface of the pathway and the water saturated flow. Despite the low speed of travel (~10 to 40 kilometres per hour (6 to 25 mph)), wet snow avalanches are capable of generating powerful destructive forces, due to the large mass, and density.

Snowmelt

Snowmelt-induced flooding of the Red River of the North in 1997

Many rivers originating in mountainous or high-latitude regions receive a significant portion of their flow from snowmelt. This often makes the river's flow highly seasonal resulting in periodic flooding during the spring months and at least in dry mountainous regions like the mountain West of the US or most of Iran and Afghanistan, very low flow for the rest of the year. In contrast, if much of the melt is from glaciated or nearly glaciated areas, the melt continues through the warm season, with peak flows occurring in mid to late summer.

Glaciers

Glaciers form where the accumulation of snow and ice exceeds ablation. The area in which an alpine glacier forms is called a cirque (corrie or cwm), a typically armchair-shaped geological feature, which collects snow and where the snowpack compacts under the weight of successive layers of accumulating snow, forming névé. Further crushing of the individual snow crystals and reduction of entrapped air in the snow turns it into glacial ice. This glacial ice will fill the cirque until it overflows through a geological weakness or an escape route, such as the gap between two mountains. When the mass of snow and ice is sufficiently thick, it begins to move due to a combination of surface slope, gravity and pressure. On steeper slopes, this can occur with as little as 15 m (50 ft) of snow-ice.

Snow Science

Scientists study snow at a wide variety of scales that include the physics of chemical bonds and clouds; the distribution, accumulation, metamorphosis, and ablation of snowpacks; and the contribution of snowmelt to river hydraulics and ground hydrology. In doing so, they employ a variety of instruments to observe and measure the phenomena studied. Their findings contribute to knowledge applied by engineers, who adapt vehicles and structures to snow, by agronomists, who address the availability of snowmelt to agriculture, and those, who design equipment for sporting activities on snow. Scientists develop and others employ snow classification systems that describe its physical properties at scales ranging from the individual crystal to the aggregated snowpack. A sub-specialty is avalanches, which are of concern to engineers and outdoors sports people, alike.

Snow science addresses how snow forms, its distribution, and processes affecting how snowpacks change over time. Scientists improve storm forecasting, study global snow cover and its effect on climate, glaciers, and water supplies around the world. The study includes physical properties of the material as it changes, bulk properties of in-place snow packs, and the aggregate properties of regions with snow cover. In doing so, they employ on-the-ground physical measurement techniques to establish ground truth and remote sensing techniques to develop understanding of snow-related processes over large areas.

Measurement and Classification

In the field snow scientists often excavate a snow pit within which to make basic measurements and observations. Observations can describe features caused by wind, water percolation, or snow unloading from trees.Water percolation into a snowpack can create flow fingers and ponding or flow along capillary barriers, which can refreeze into horizontal and vertical solid ice formations within the snowpack. Among the measurements of the properties of snowpacks that the *International Classification for Seasonal Snow on the Ground* includes are: snow height, snow water equivalent, snow strength, and extent of snow cover. Each has a designation with code and detailed description. The classification extends the prior classifications of Nakaya and his successors to related types of precipitation and are quoted in the following table:

Snow pit on the surface of a glacier, profiling snow properties where the snow becomes increasingly dense with depth as it metamorphoses towards ice

Frozen precipitation particles, related to snow crystals		
Subclass	Shape	Physical process
Graupel	Heavily rimed particles, spherical, conical, hexagonal or irregular in shape	Heavy riming of particles by accretion of supercooled water droplets
Hail	Laminar internal structure, translucent or milky glazed surface	Growth by accretion of supercooled water, size: >5 mm

Ice pellets	Transparent, mostly small spheroids	Freezing of raindrops or refreezing of largely melted snow crystals or snowflakes (sleet).Graupel or snow pellets encased in thin ice layer (small hail). Size: both 5 mm
Rime	Irregular deposits or longer cones and needles pointing into the wind	Accretion of small, supercooled fog droplets frozen in place. Thin breakable crust forms on snow surface if process continues long enough.

All are formed in cloud, except for rime, which forms on objects exposed to supercooled moisture.

It also has a more extensive classification of deposited snow than those that pertain to airborne snow. The categories include both natural and man-made snow types, descriptions of snow crystals as they metamorphose and melt, the development of hoar frost in the snow pack and the formation of ice therein. Each such layer of a snowpack differs from the adjacent layers by one or more characteristics that describe its microstructure or density, which together define the snow type, and other physical properties. Thus, at any one time, the type and state of the snow forming a layer have to be defined because its physical and mechanical properties depend on them. Physical properties include microstructure, grain size and shape, snow density, liquid water content, and temperature.

Satellite Data

Remote sensing of snowpacks with satellites and other platforms typically includes multi-spectral collection of imagery. Multi-faceted interpretation of the data obtained allows inferences about what is observed. The science behind these remote observations has been verified with ground-truth studies of the actual conditions.

Satellite observations record a decrease in snow-covered areas since the 1960s, when satellite observations began. In some regions such as China, a trend of increasing snow cover was observed from 1978 to 2006. These changes are attributed to global climate change, which may lead to earlier melting and less coverage area. However, in some areas there may be an increase in snow depth because of higher temperatures for latitudes north of 40°. For the Northern Hemisphere as a whole the mean monthly snow-cover extent has been decreasing by 1.3% per decade.

The most frequently used methods to map and measure snow extent, snow depth and snow water equivalent employ multiple inputs on the visible–infrared spectrum to deduce the presence and properties of snow. The National Snow and Ice Data Center (NSIDC) uses the reflectance of visible and infrared radiation to calculate a normalized difference snow index, which is a ratio of radiation parameters that can distinguish between clouds and snow. Other researchers have developed decision trees, employing the available data to make more accurate assessments. One challenge to this assessment is where snow cover is patchy, for example during periods of accumulation or ablation and also in forested areas. Cloud cover inhibits optical sensing of surface reflectance, which has led to other methods for estimating ground conditions underneath clouds. For hydrological models, it is important to have continuous information about the snow cover. Passive microwave sensors are especially valuable for temporal and spatial continuity because they can map the surface beneath clouds and in darkness. When combined with reflective measurements, passive microwave sensing greatly extends the inferences possible about the snowpack.

Models

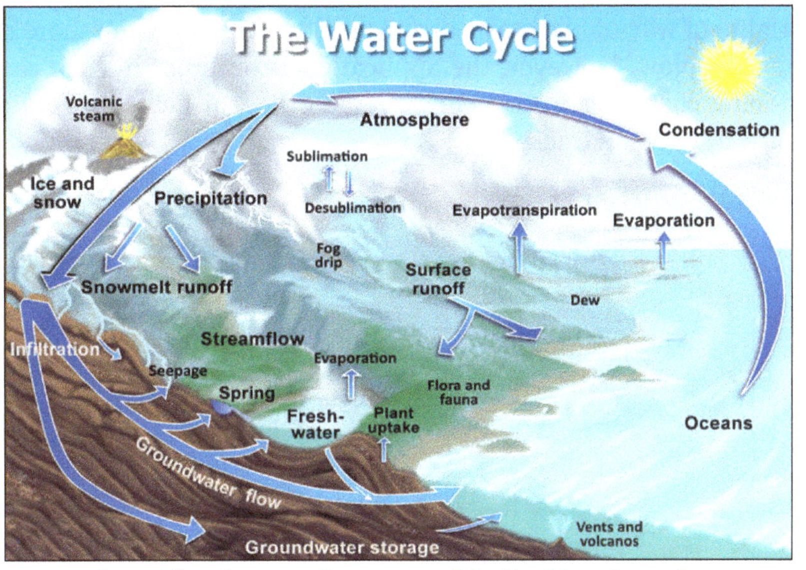

Snowfall and snowmelt are parts of the Earth's water cycle

Snow science often leads to predictive models that include snow deposition, snow melt, and snow hydrology—elements of the Earth's water cycle—which help describe global climate change.

Global climate change models (GCMs) incorporate snow as a factor in their calculations. Some important aspects of snow cover include its albedo (reflectivity of incident radiation, including light) and insulating qualities, which slow the rate of seasonal melting of sea ice. As of 2011, the melt phase of GCM snow models were thought to perform poorly in regions with complex factors that regulate snow melt, such as vegetation cover and terrain. These models typically derive snow water equivalent (SWE) in some manner from satellite observations of snow cover. The *International Classification for Seasonal Snow on the Ground* defines SWE as "the depth of water that would result if the mass of snow melted completely".

Given the importance of snowmelt to agriculture, hydrological runoff models that include snow in their predictions address the phases of accumulating snowpack, melting processes, and distribution of the meltwater through stream networks and into the groundwater. Key to describing the melting processes are solar heat flux, ambient temperature, wind, and precipitation. Initial snowmelt models used a degree-day approach that emphasized the temperature difference between the air and the snowpack to compute snow water equivalent, SWE. More recent models use an energy balance approach that take into account the following factors to compute Q_m, the energy available for melt. This requires measurement of an array of snowpack and environmental factors to compute six heat flow mechanisms that contribute to Q_m.

Effects on Human Activity

Snow affects human activity in four major areas, transportation, agriculture, structures, and sports. Most transportation modes are impeded by snow on the travel surface. Agriculture often relies on snow as a source of seasonal moisture. Structures may fail under snow loads. Humans find a wide variety of recreational activities in snowy landscapes.

Transportation

Snow affects the rights of way of highways, airfields and railroads. They share a common tool for clearing snow, the snowplow. However, the application is different in each case—whereas roadways employ anti-icing chemicals to prevent bonding of ice, airfields may not; railroads rely on abrasives to enhance traction on tracks.

Highway

Traffic stranded in a 2011 Chicago snowstorm

In the late 20th century, an estimated $2 billion was spent annually in North America on roadway winter maintenance, owing to snow and other winter weather events, according to a 1994 report by Kuemmel. The study surveyed the practices of jurisdictions within 44 US states and nine Canadian provinces. It assessed the policies, practices, and equipment used for winter maintenance. It found similar practices and progress to be prevalent in Europe.

Winter conditions on Ontario Highway 401 in Toronto due to a snowsquall

The dominant effect of snow on vehicle contact with the road is diminished friction. This can be improved with the use of snow tires, which have a tread designed to compact snow in a manner that enhances traction. However, the key to maintaining a roadway that can accommodate traffic during and after a snow event is an effective anti-icing program that employs both chemicals and plowing. The FHWA *Manual of Practice for an Effective Anti-icing Program* emphasizes "anti-icing" procedures that prevent the bonding of snow and ice to the road. Key aspects of the practice include: understanding anti-icing in light of the level of service to be achieved on a given roadway, the climatic conditions to be encountered, and the different roles of deicing, anti-icing, and abrasive materials and applications, and employing anti-icing "toolboxes", one for operations, one for decision-making and another for personnel. The elements to the toolboxes are:

- *Operations* – Addresses the application of solid and liquid chemicals, using various techniques, including prewetting of chloride-salts. It also addresses plowing capability, including types of snowplows and blades used.
- *Decision-making* – Combines weather forecast information with road information to assess the upcoming needs for application of assets and the evaluation of treatment effectiveness with operations underway.
- *Personnel* – Addresses training and deployment of staff to effectively execute the anti-icing program, using the appropriate materials, equipment and procedures.

The manual offers matrices that address different types of snow and the rate of snowfall to tailor applications appropriately and efficiently.

Snow fences, constructed upwind of roadways control snow drifting by causing windblown, drifting snow to accumulate in a desired place. They are also used on railways. Additionally, farmers and ranchers use snow fences to create drifts in basins for a ready supply of water in the spring.

Aviation

In order to keep airports open during winter storms, runways and taxiways require snow removal. Unlike roadways, where chloride chemical treatment is common to prevent snow from bonding to the pavement surface, such chemicals are typically banned from airports because of their strong corrosive effect on aluminum aircraft. Consequently, mechanical brushes are often used to complement the action of snow plows. Given the width of runways on airfields that handle large aircraft, vehicles with large plow blades, an echelon of plow vehicles or rotary snowplows are used to clear snow on runways and taxiways. Terminal aprons may require 6 hectares (15 acres) or more to be cleared.

Deicing an aircraft during a snow event

Properly equipped aircraft are able to fly through snowstorms under Instrument flight rules. Prior to takeoff, during snowstorms they require deicing fluid to prevent accumulation and freezing of snow and other precipitation on wings and fuselages, which may compromise the safety of the aircraft and its occupants. In flight, aircraft rely on a variety of mechanisms to avoid rime and other types of icing in clouds, these include pulsing pneumatic boots, electro-thermal areas that generate heat, and fluid deicers that bleed onto the surface.

Rail

Railroads have traditionally employed two types of snow plows for clearing track, the wedge plow, which casts snow to both sides, and the rotary snowplow, which is suited for addressing heavy snowfall and casting snow far to one side or the other. Prior to the invention of the rotary snowplow ca. 1865, it required multiple locomotives to drive a wedge plow through deep snow. Subsequent to clearing the track with such plows, a "flanger" is used to clear snow from between the rails that are below the reach of the other types of plow. Where icing may affect the steel-to-steel contact of locomotive wheels on track, abrasives (typically sand) have been used to provide traction on steeper uphills.

Railroads employ snow sheds—structures that cover the track—to prevent the accumulation of heavy snow or avalanches to cover tracks in snowy mountainous areas, such as the Alps and the Rocky Mountains.

Snowplows for Different Transportation Modes

Snow Roads and Runways

Snow can be compacted to form a snow road and be part of a winter road route for vehicles to access isolated communities or construction projects during the winter. Snow can also be used to provide the supporting structure and surface for a runway, as with the Phoenix Airfield in Antarctica. The snow-compacted runway is designed to withstand approximately 60 wheeled flights of heavy-lift military aircraft a year.

Trucks plowing snow on a highway in Missouri

Swiss low-profile, train-mounted snowplow

Agriculture

Snowfall can be beneficial to agriculture by serving as a thermal insulator, conserving the heat of the Earth and protecting crops from subfreezing weather. Some agricultural areas depend on an accumulation of snow during winter that will melt gradually in spring, providing water for crop growth, both directly and via runoff through streams and rivers, which supply irrigation canals.

The following are examples of rivers that rely on meltwater from glaciers or seasonal snowpack as an important part of their flow on which irrigation depends: the Ganges, many of whose tributaries rise in the Himalayas and which provide much irrigation in northeast India, the Indus River, which rises in Tibet and provides irrigation water to Pakistan from rapidly retreating Tibetan glaciers, and the Colorado River, which receives much of its water from seasonal snowpack in the Rocky Mountains and provides irrigation water to some 4 million acres (1.6 million hectares).

Satellite view of the Indus River, showing snow in the Himalayas, which feeds it, and green areas that draw on it for irrigation

Structures

Snow is an important consideration for loads on structures. To address these, European countries employ *Eurocode 1: Actions on structures - Part 1-3: General actions - Snow loads.* In North America, ASCE *Minimum Design Loads for Buildings and Other Structures* gives guidance on snow loads. Both standards employ methods that translate maximum expected ground snow loads onto design loads for roofs.

Snow accumulation on building roofs

Roofs

Snow loads and icings are two principal issues for roofs. Snow loads are related to the climate in which a structure is sited. Icings are usually a result of the building or structure generating heat that melts the snow that is on it.

Snow loads – The *Minimum Design Loads for Buildings and Other Structures* gives guidance on how to translate the following factors into roof snow loads:

- Ground snow loads.
- Exposure of the roof.
- Thermal properties of the roof.
- Shape of the roof.
- Drifting.
- Importance of the building.

It gives tables for ground snow loads by region and a methodology for computing ground snow loads that may vary with elevation from nearby, measured values. The *Eurocode 1* uses similar methodologies, starting with ground snow loads that are tabulated for portions of Europe.

Icings – Roofs must also be designed to avoid ice dams, which result from meltwater running under the snow on the roof and freezing at the eave. Ice dams on roofs form when accumulated snow on a sloping roof melts and flows down the roof, under the insulating blanket of snow, until it reaches below freezing temperature air, typically at the eaves. When the meltwater reaches the freezing air, ice accumulates, forming a dam, and snow that melts later cannot drain properly through the dam. Ice dams may result in damaged building materials or in damage or injury when the ice dam falls off or from attempts to remove ice dams. The melting results from heat passing through the roof under the highly insulating layer of snow.

Icings resulting from meltwater at the bottom of the snow pack on the roof, flowing and refreezing at the eave as icicles and from leaking into the wall via an ice dam

Utility Lines

In areas with trees, utility distribution lines on poles are less susceptible to snow loads than they are subject to damage from trees falling on them, felled by heavy, wet snow. Elsewhere, snow can accrete on power lines as "sleeves" of rime ice. Engineers design for such loads, which are measured in kg/m (lb/ft) and power companies have forecasting systems that anticipate types of weather that may cause such accretions. Rime ice may be removed manually or by creating a sufficient short circuit in the affected segment of power lines to melt the accretions.

Sports and Recreation

Snow figures into many winter sports and forms of recreation, including skiing and sledding. Common examples include cross-country skiing, Alpine skiing, snowboarding, snowshoeing, and snowmobiling. The design of the equipment used, typically relies on the bearing strength of snow, as with skis or snowboards and contends with the coefficient of friction of snow to allow sliding, often enhance by ski waxes.

Alpine skiing

Skiing is by far the largest form of winter recreation. As of 1994, of the estimated 65–75 million skiers worldwide, there were approximately 55 million who engaged in Alpine skiing, the rest engaged in cross-country skiing. Approximately 30 million skiers (of all kinds) were in Europe, 15 million in the US, and 14 million in Japan. As of 1996, there were reportedly 4,500 ski areas, operating 26,000 ski lifts and enjoying 390 million skier visits per year. The preponderant region for downhill skiing was Europe, followed by Japan and the US.

Increasingly, ski resorts are relying on snowmaking, the production of snow by forcing water and pressurized air through a snow gun on ski slopes. Snowmaking is mainly used to supplement natural snow at ski resorts. This allows them to improve the reliability of their snow cover and to extend their ski seasons from late autumn to early spring. The production of snow requires low temperatures. The threshold temperature for snowmaking increases as humidity decreases. Wet-bulb temperature is used as a metric since it takes air temperature and relative humidity into account. Snowmaking is a relatively expensive process in its energy consumption, thereby limiting its use.

Ski wax enhances the ability of a ski or other runner to slide over snow, which depends on both the properties of the snow and the ski to result in an optimum amount of lubrication from melting the snow by friction with the ski—too little and the ski interacts with solid snow crystals, too much and capillary attraction of meltwater retards the ski. Before a ski can slide, it must overcome the maximum value static friction. Kinetic (or dynamic) friction occurs when the ski is moving over the snow.

Warfare

Snow affects warfare conducted in winter, alpine environments or at high latitudes. The main factors are *impaired visibility* for acquiring targets during falling snow, *enhanced visibility* of targets against snowy backgrounds for targeting, and mobility for both mechanized and infantry troops. Snowfall can severely inhibit the logistics of supplying troops, as well. Snow can also provide cover and fortification against small-arms fire. Noted winter warfare campaigns where snow and other factors affected the operations include:

- The French invasion of Russia, where poor traction conditions for ill-shod horses made it difficult for supply wagons to keep up with troops. That campaign was also strongly affected by cold, whereby the retreating army reached Neman River in December 1812 with only 10,000 of the 420,000 that had set out to invade Russia in June of the same year.
- The Winter War, an attempt by the Soviet Union to take territory in Finland in late 1939 demonstrated superior winter tactics of the Finnish Army, regarding over-snow mobility, camouflage, and use of the terrain.
- The Battle of the Bulge, a German counteroffensive during World War II, starting December 16, 1944, was marked by heavy snowstorms that hampered allied air support for ground troops, but also impaired German attempts to supply their front lines. On the Eastern Front with the Nazi invasion of Russia in 1941, Operation Barbarossa, both Russian and German soldiers had to endure terrible conditions during the Russian winter. While use of ski infantry was common in the Red Army, Germany formed only one division for movement on skis.
- The Korean War which lasted from June 25, 1950, until an armistice on July 27, 1953, began when North Korea invaded South Korea. Much of the fighting occurred during winter conditions, involving snow, notably during the Battle of Chosin Reservoir, which was a stark example of cold affecting military operations, especially vehicles and weapons.

Effects on Ecosystems

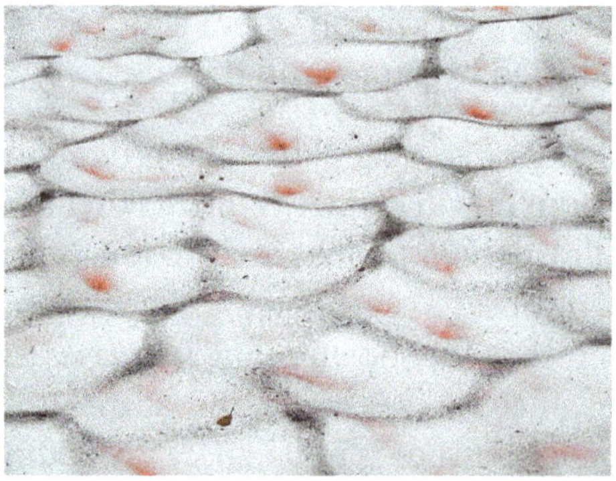

Algae, *Chlamydomonas nivalis*, that thrive in snow form red areas in the suncups on this snow surface

Both plant and animal life endemic to snow-bound areas develop ways to adapt. Among the adaptive mechanisms for plants are dormancy, seasonal dieback, survival of seeds; and for animals are hibernation, insulation, anti-freeze chemistry, storing food, drawing on reserves from within the body, and clustering for mutual heat.

Plant Life

Snow interacts with vegetation in two principal ways, vegetation can influence the deposition and retention of snow and, conversely, the presence of snow can affect the distribution and growth of vegetation. Tree branches, especially of conifers intercept falling snow and prevent accumulation on the ground. Snow suspended in trees ablates more rapidly than that on the ground, owing to its greater exposure to sun and air movement. Trees and other plants can also promote snow retention on the ground, which would otherwise be blown elsewhere or melted by the sun. Snow affects vegetation in several ways, the presence of stored water can promote growth, yet the annual onset of growth is dependent on the departure of the snowpack for those plants that are buried beneath it. Furthermore, avalanches and erosion from snowmelt can scour terrain of vegetation.

Animal Life

Arctic fox, a predator of smaller animals that live beneath the snow

Snow supports a wide variety of animals both on the surface and beneath. Many invertebrates thrive in snow, including spiders, wasps, beetles, snow scorpionflys and springtails. Such arthropods are typically active at temperatures down to −5 °C (23 °F). Invertebrates fall into two groups, regarding surviving subfreezing temperatures: freezing resistant and those that avoid freezing because they are freeze-sensitive. The first group may be cold hardy owing to the ability to produce antifreeze agents in their body fluids that allows survival of long exposure to sub-freezing conditions. Some organisms fast during the winter, which expels freezing-sensitive contents from their digestive tracts. The ability to survive the absence of oxygen in ice is an additional survival mechanism.

Small vertebrates are active beneath the snow. Among vertebrates, alpine salamanders are active in snow at temperatures as low as −8 °C (18 °F); they burrow to the surface in springtime and lay their eggs in melt ponds. Among mammals, those that remain active are typically smaller than 250 grams (8.8 oz). Omnivores are more likely to enter a torpor or be hibernators, whereas herbivores are more likely to maintain food caches beneath the snow. Voles store up to 3 kilograms (6.6 lb) of food and pikas up to 20 kilograms (44 lb). Voles also huddle in communal nests to benefit from one another's warmth. On the surface, wolves, coyotes, foxes, lynx, and weasels rely on these sub-surface dwellers for food and often dive into the snowpack to find them.

TEMPERATURE

Temperature is a physical quantity expressing hot and cold. It is measured with a thermometer calibrated in one or more temperature scales. The most commonly used scales are the Celsius scale (formerly called *centigrade*) (denoted °C), Fahrenheit scale (denoted °F), and Kelvin scale (denoted K). The kelvin (the word is spelled with a lower-case k) is the unit of temperature in the International System of Units (SI). The Kelvin scale is widely used in science and technology.

Theoretically, the coldest a system can be is when its temperature is absolute zero, at which point the thermal motion in matter would be zero. However, an actual physical system or object can never attain a temperature of absolute zero. Absolute zero is denoted as 0 K on the Kelvin scale, −273.15 °C on the Celsius scale, and −459.67 °F on the Fahrenheit scale.

For an ideal gas, temperature is proportional to the average kinetic energy of the random microscopic motions of the constituent microscopic particles. This is now the basis of the definition of the magnitude of the kelvin.

Temperature is important in all fields of natural science, including physics, chemistry, Earth science, medicine, and biology, as well as most aspects of daily life.

Effects

Many physical processes are affected by temperature, such as:

- Physical properties of materials including the phase (solid, liquid, gaseous or plasma), density, solubility, vapor pressure, electrical conductivity.
- Rate and extent to which chemical reactions occur.
- The amount and properties of thermal radiation emitted from the surface of an object.
- Speed of sound is a function of the square root of the absolute temperature.

Scales

Temperature scales differ in two ways: the point chosen as zero degrees, and the magnitudes of incremental units or degrees on the scale.

The Celsius scale (°C) is used for common temperature measurements in most of the world. It is an empirical scale that was developed by a historical progress, which led to its zero point 0 °C being defined by the freezing point of water, and additional degrees defined so that 100 °C was the boiling point of water, both at sea-level atmospheric pressure. Because of the 100-degree interval, it was called a centigrade scale. Since the standardization of the kelvin in the International System of Units, it has subsequently been redefined in terms of the equivalent fixing points on the Kelvin scale, and so that a temperature increment of one degree Celsius is the same as an increment of one kelvin, though they differ by an additive offset of approximately 273.15.

The United States commonly uses the Fahrenheit scale, on which water freezes at 32 °F and boils at 212 °F at sea-level atmospheric pressure.

Many scientific measurements use the Kelvin temperature scale (unit symbol: K), named in honor of the Scots-Irish physicist who first defined it. It is a thermodynamic or absolute temperature scale. Its zero point, 0 K, is defined to coincide with the coldest conceivable temperature (called absolute zero). Its degrees are defined through particle kinetic theory.

The process of cooling involves removing energy from a system. When no more energy can be removed, the system is at absolute zero, though this cannot be achieved experimentally. Absolute zero is the null point of the thermodynamic temperature scale, also called absolute temperature. If it were possible to cool a system to absolute zero, all classical motion of its particles would cease and they would be at complete rest in this classical sense. Microscopically in the description of quantum mechanics, however, matter still has zero-point energy even at absolute zero, because of the uncertainty principle. Such zero-point energy is not considered "heat-driven" or "thermal" motion and does not enter into the definition of thermodynamic, or absolute, temperature.

Until May 2019, the International System of Units (SI) used to define a scale and unit for the kelvin or thermodynamic temperature by using the reliably reproducible temperature of the triple point of water as a second reference point (the first reference point being 0 K at absolute zero). The triple point is a singular state with its own unique and invariant temperature and pressure, along with, for a fixed mass of water in a vessel of fixed volume, an autonomically and stably self-determining partition into three mutually contacting phases, vapour, liquid, and solid, dynamically depending only on the total internal energy of the mass of water. Historically, the triple point temperature of water was defined to be exactly at 273.16 units of the measurement increment.

Dew Point

The dew point is the temperature to which air must be cooled to become saturated with water vapor. When further cooled, the airborne water vapor will condense to form liquid water (dew). When air cools to its dew point through contact with a surface that is colder than the air, water will condense on the surface. When the temperature is below the freezing point of water, the dew point is called the frost point, as frost is formed rather than dew. The measurement of the dew point is related to humidity. A higher dew point means there is more moisture in the air.

Humidity

If all the other factors influencing humidity remain constant, at ground level the relative humidity rises as the temperature falls; this is because less vapor is needed to saturate the air. In normal conditions, the dew point temperature will not be greater than the air temperature, since relative humidity cannot exceed 100%.

In technical terms, the dew point is the temperature at which the water vapor in a sample of air at constant barometric pressure condenses into liquid water at the same rate at which it evaporates. At temperatures below the dew point, the rate of condensation will be greater than that of evaporation, forming more liquid water. The condensed water is called dew when it forms on a solid surface, or frost if it freezes. In the air, the condensed water is called either fog or a cloud, depending on its altitude when it forms. If the temperature is below the dew point, the vapor is called supersaturated. This can happen if there are not enough particles in the air to act as condensation nuclei.

A high relative humidity implies that the dew point is closer to the current air temperature. A relative humidity of 100% indicates the dew point is equal to the current temperature and that the air is maximally saturated with water. When the moisture content remains constant and temperature increases, relative humidity decreases, but the dew point remains constant.

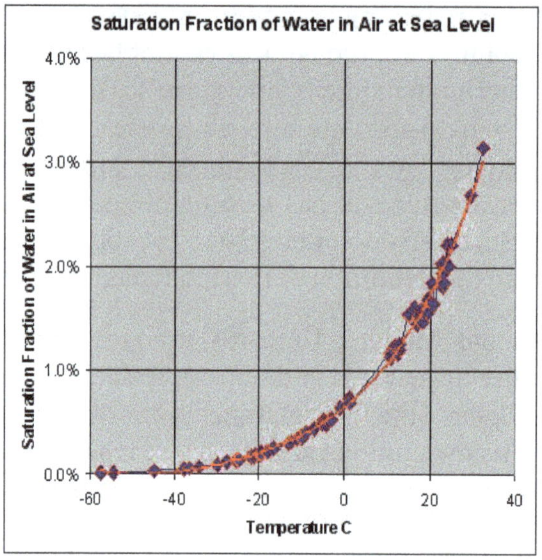

This graph shows the maximum percentage, by mass, of water vapor that air at sea-level pressure across a range of temperatures can contain. For a lower ambient pressure, a curve has to be drawn above the current curve. A higher ambient pressure yields a curve under the current curve

General aviation pilots use dew point data to calculate the likelihood of carburetor icing and fog, and to estimate the height of a cumuliform cloud base.

Increasing the barometric pressure increases the dew point. This means that, if the pressure increases, the mass of water vapor in the air must be reduced in order to maintain the same dew point. For example, consider New York (33 ft or 10 m elevation) and Denver (5,280 ft or 1,610 m elevation). Because Denver is at a higher elevation than New York, it will tend to have a lower barometric pressure. This means that if the dew point and temperature in both cities are the same, the amount of water vapor in the air will be greater in Denver.

Relationship to Human Comfort

When the air temperature is high, the human body uses the evaporation of sweat to cool down, with the cooling effect directly related to how fast the perspiration evaporates. The rate at which perspiration can evaporate depends on how much moisture is in the air and how much moisture the air can hold. If the air is already saturated with moisture, perspiration will not evaporate. The body's thermoregulation will produce perspiration in an effort to keep the body at its normal temperature even when the rate it is producing sweat exceeds the evaporation rate, so one can become coated with sweat on humid days even without generating additional body heat (such as by exercising).

As the air surrounding one's body is warmed by body heat, it will rise and be replaced with other air. If air is moved away from one's body with a natural breeze or a fan, sweat will evaporate faster, making perspiration more effective at cooling the body. The more unevaporated perspiration, the greater the discomfort.

A wet bulb thermometer also uses evaporative cooling, so it provides a good measure for use in evaluating comfort level.

Discomfort also exists when the dew point is very low (below around −30 °C or −22 °F). The drier air can cause skin to crack and become irritated more easily. It will also dry out the airways. The US Occupational Safety and Health Administration recommends indoor air be maintained at 20–24.5 °C (68–76 °F) with a 20–60% relative humidity, equivalent to a dew point of −4.5 to 15.5 °C (24 to 60 °F).

Lower dew points, less than 10 °C (50 °F), correlate with lower ambient temperatures and the body requires less cooling. A lower dew point can go along with a high temperature only at extremely low relative humidity, allowing for relatively effective cooling.

People inhabiting tropical and subtropical climates acclimatize somewhat to higher dew points. Thus, a resident of Singapore or Miami, for example, might have a higher threshold for discomfort than a resident of a temperate climate like London or Chicago. People accustomed to temperate climates often begin to feel uncomfortable when the dew point gets above 15 °C (59 °F), while others might find dew points up to 18 °C (64 °F) comfortable. Most inhabitants of temperate areas will consider dew points above 21 °C (70 °F) oppressive and tropical-like, while inhabitants of hot and humid areas may not find this uncomfortable. Thermal comfort depends not just on physical environmental factors, but also on psychological factors.

Dew point		Relative humidity at 32 °C (90 °F)
Over 26 °C	Over 80 °F	73% and higher
24–26 °C	75–80 °F	62–72%
21–24 °C	70–74 °F	52–61%
18–21 °C	65–69 °F	44–51%
16–18 °C	60–64 °F	37–43%
13–16 °C	55–59 °F	31–36%
10–12 °C	50–54 °F	26–30%
Under 10 °C	Under 50 °F	25% and lower

Measurement

Devices called hygrometers are used to measure dew point over a wide range of temperatures. These devices consist of a polished metal mirror which is cooled as air is passed over it. The temperature at which dew forms is, by definition, the dew point. Manual devices of this sort can be used to calibrate other types of humidity sensors, and automatic sensors may be used in a control loop with a humidifier or dehumidifier to control the dew point of the air in a building or in a smaller space for a manufacturing process.

Calculating the Dew Point

A well-known approximation used to calculate the dew point, T_{dp}, given just the actual ("dry bulb") air temperature, T (in degrees Celsius) and relative humidity (in percent), RH, is the Magnus formula:

$$\gamma(T,\mathrm{RH}) = \ln\left(\frac{\mathrm{RH}}{100}\right) + \frac{bT}{c+T};$$

$$T_{dp} = \frac{c\gamma(T,\mathrm{RH})}{b-\gamma(T,\mathrm{RH})};$$

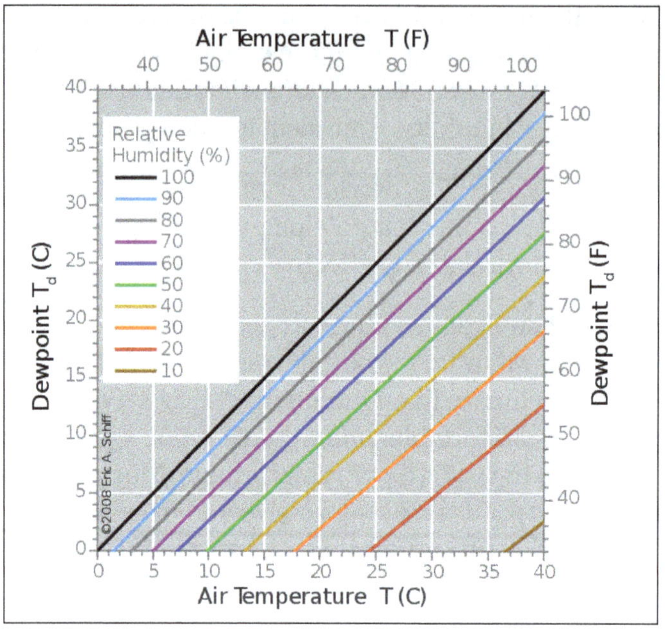

Graph of the dependence of the dew point upon air temperature for several levels of relative humidity

The more complete formulation and origin of this approximation involves the interrelated saturated water vapor pressure (in units of millibars, also called hectopascals) at T, $P_s(T)$, and the actual vapor pressure (also in units of millibars), $P_a(T)$, which can be either found with RH or approximated with the barometric pressure (in millibars), BP_{mbar}, and "wet-bulb" temperature, T_w is (unless declared otherwise, all temperatures are expressed in degrees Celsius):

$$P_s(T) = \frac{100}{\mathrm{RH}} P_a(T) = ae^{\frac{bT}{c+T}};$$

$$P_a(T) = \frac{\mathrm{RH}}{100} P_s(T) = ae^{\gamma(T,\mathrm{RH})}$$

$$\approx P_s(T_w) - BP_{\mathrm{mbar}}\, 0.00066(1+0.00115T_w)(T-T_w);$$

$$T_{dp} = \frac{c \ln \frac{P_a(T)}{a}}{b - \ln \frac{P_a(T)}{a}};$$

For greater accuracy, $P_s(T)$ (and therefore $\gamma(T, \mathrm{RH})$) can be enhanced, using part of the *Bögel modification*, also known as the Arden Buck equation, which adds a fourth constant d:

$$P_{s,m}(T) = ae^{\left(b-\frac{T}{d}\right)\left(\frac{T}{c+T}\right)};$$

$$\gamma_m(T,\mathrm{RH}) = \ln\left(\frac{\mathrm{RH}}{100} e^{\left(b-\frac{T}{d}\right)\left(\frac{T}{c+T}\right)}\right);$$

$$T_{dp} = \frac{c\gamma_m(T,\mathrm{RH})}{b-\gamma_m(T,\mathrm{RH})};$$

where,

- $a = 6.1121$ mbar, $b = 18.678$, $c = 257.14$ °C, $d = 234.5$ °C.

There are several different constant sets in use. The ones used in NOAA's presentation are taken from a 1980 paper by David Bolton in the *Monthly Weather Review*:

- $a = 6.112$ mbar, $b = 17.67$, $c = 243.5$ °C.

These valuations provide a maximum error of 0.1%, for -30 °C $\leq T \leq 35$ °C and 1% < RH < 100%. Also noteworthy is the Sonntag1990,

- $a = 6.112$ mbar, $b = 17.62$, $c = 243.12$ °C; for -45 °C $\leq T \leq 60$ °C (error ±0.35 °C).

Another common set of values originates from the 1974 *Psychrometry and Psychrometric Charts*, as presented by *Paroscientific*,

- $a = 6.105$ mbar, $b = 17.27$, $c = 237.7$ °C; for 0 °C $\leq T \leq 60$ °C (error ±0.4 °C).

Arden Buck presents several different valuation sets, with different minimum accuracies for different temperature ranges. Two particular sets provide a range of −40 °C to +50 °C between the two, with even greater minimum accuracy than all of the other, above sets (maximum error at extremes of temperature range):

- $a = 6.1121$ mbar, $b = 17.368$, $c = 238.88$ °C; for 0 °C $\leq T \leq 50$ °C (error ≤ 0.05%).
- $a = 6.1121$ mbar, $b = 17.966$, $c = 247.15$ °C; for -40 °C $\leq T \leq 0$ °C (error ≤ 0.06%).

Simple Approximation

There is also a very simple approximation that allows conversion between the dew point, temperature, and relative humidity. This approach is accurate to within about ±1 °C as long as the relative humidity is above 50%:

$$\begin{aligned} T_{\text{dp}} &\approx T - \frac{100 - \text{RH}}{5}; \\ \text{RH} &\approx 100 - 5(T - T_{\text{dp}}); \end{aligned}$$

This can be expressed as a simple rule of thumb:

For every 1 °C difference in the dew point and dry bulb temperatures, the relative humidity decreases by 5%, starting with RH = 100% when the dew point equals the dry bulb temperature.

For temperatures in degrees Fahrenheit, these approximations work out to:

$$\begin{aligned} T_{\text{dp},^\circ\text{F}} &\approx T_{^\circ\text{F}} - \frac{9}{25}(100 - \text{RH}); \\ \text{RH} &\approx 100 - \frac{25}{9}\left(T_{^\circ\text{F}} - T_{\text{dp},^\circ\text{F}}\right); \end{aligned}$$

For example, a relative humidity of 100% means dew point is the same as air temp. For 90% RH, dew point is 3 °F lower than air temperature. For every 10 percent lower, dew point drops 3 °F.

Frost Point

The frost point is similar to the dew point in that it is the temperature to which a given parcel of humid air must be cooled, at constant atmospheric pressure, for water vapor to be deposited on a surface as ice crystals without undergoing the liquid phase (compare with sublimation). The frost point for a given parcel of air is always higher than the dew point, as the stronger bonding between water molecules on the surface of ice requires higher temperature to break.

Sea Surface Temperature

Sea surface temperature is a key climate and weather measurement obtained by satellite microwave radiometers, infrared (IR) radiometers, in situ moored and drifting buoys, and ships of opportunity. Different instruments measure the temperature at different depths. For instance, most buoys have sensors located at about 1 meter depth, or placed at regular intervals along a tether line. Sea surface temperatures, when measured from space, represent a depth that is related to the frequency of the satellite instrument. For example, IR instruments measure a depth of about 20 micrometers, while microwave radiometers measure a depth of a few millimeters. The microwave (MW) optimally-interpolated (OI) SSTs are designed to represent a foundation SST at a depth of ~1 meter, or temperatures just below the diurnal layer.

These measurements can be combined in various ways to create daily spatially-complete global SST maps used for weather prediction, ocean forecasts, and in coastal applications such as fisheries forecasts, pollution monitoring, and tourism. SST maps are also widely used by oceanographers, meteorologists, and climate scientists for scientific research. Prior to 1997, SSTs were only available globally from IR satellite retrievals, but with the launch of TMI, microwave retrievals became possible. While IR SSTs have a higher resolution than microwave SSTs (1 – 4 km for IR as compared to 25 km for microwave), the IR retrieval is prevented by clouds giving microwave SSTs improved coverage since SST can be measured through clouds. This has proven especially important in tropical cyclone forecasting as the clouds surrounding a cyclone prevented adequate SST measurements until microwave instruments became available in 1998.

Microwave Measurement of Sea Surface Temperature

At frequencies between 4 and 11 GHz, the vertically polarized microwave brightness temperature of ocean areas has an appreciable sensitivity to SST. In addition to SST, brightness temperature depends on the sea-surface roughness and on the atmospheric temperature and moisture profile. Fortunately, the spectral and polarimetric signatures of the surface-roughness and the atmosphere are quite distinct from the SST signature, and the influence of these effects can be removed given simultaneous measurements at multiple frequencies and polarizations. The microwave instruments TMI, AMSR-E, AMSR2, WindSat and GMI all measure multiple frequencies that are more than sufficient to remove the surface-roughness and atmospheric effects. Sea-surface roughness, which is tightly correlated with the local wind, is usually parameterized in terms of the near-surface wind speed and direction. The additional 7 GHz channel present on AMSR-E and AMSR2 but not TMI, provides improved estimates of sea-surface roughness and improved accuracy for SSTs less than 12°C. All channels are used to simultaneously retrieve SST, wind speed, columnar water vapor, cloud liquid water, and rain rate. SST retrieval is prevented only in regions with sun-glitter, rain, or proximity to land. Since only a small number of retrievals are unsuccessful, almost complete global coverage is achieved daily. Any errors in retrieved wind speed, water vapor, or cloud liquid water can result in errors in the retrieved SSTs.

The through-cloud capabilities of microwave radiometers provide a valuable picture of global sea surface temperature (SST). To utilize this, scientists at RSS have calculated a daily, Optimally Interpolated (OI) SST product at quarter degree (~25 kilometer) resolution. This product is ideal for research activities in which a complete, daily SST map is more desirable than one with missing data due to orbital gaps or environmental conditions precluding SST retrieval.

RSS Sea Surface Temperature Products

Daily, global optimally interpolated (OI) ssts: To provide users with complete maps of SST, we have two optimally interpolated SST products that merge multiple satellites (when available) into globally complete, daily, SST maps.

Microwave satellite sst observations, both swath and gridded, available as ghrsst netcdf files: The SST values from the individual satellite binary files are extracted.

Individual radiometer gridded binary data file: Sst is one of the ocean measurements derived from microwave data processing for the instruments that have the lower frequency channels (6-7 GHz and/or 11 GHz). We include SST in the RSS satellite gridded binary data files for TMI, AMSR-E, AMSR2, WindSat, and GMI. These SST are not diurnally corrected and represent the value at the time provided in the file. The table below shows the period of operation for each instrument.

Instrument	Period of Operation	Version
TMI	1997 - 2015	V7.1
AMSR-E	2002 - 2011	V7
WindSat	2003 - present	V7.0.1
AMSR2	2012 - present	V7.1
GMI	2014 - present	V8.1

Sea surface temperatures have a large influence on climate and weather. For example, every 3 to 7 years a wide swath of the Pacific Ocean along the equator warms by 2 to 3 degrees Celsius. This warming is a hallmark of the climate pattern El Niño, which changes rainfall patterns around the globe, causing heavy rainfall in the southern United States and severe drought in Australia, Indonesia, and southern Asia. On a smaller scale, ocean temperatures influence the development of tropical cyclones (hurricanes and typhoons), which draw energy from warm ocean waters to form and intensify.

These sea surface temperature maps are based on observations by the Moderate Resolution Imaging Spectroradiometer (MODIS) on NASA's Aqua satellite. The satellite measures the temperature of the top millimeter of the ocean surface. In this map, the coolest waters appear in blue (approximately -2 degrees Celsius), and the warmest temperatures appear in pink-yellow (35 degrees Celsius). Landmasses and the large area of sea ice around Antarctica appear in shades of gray, indicating no data were collected.

The most obvious pattern shown in the time series is the year-round difference in sea surface temperatures between equatorial regions and the poles. Various warm and cool currents stand out even in monthly averages of sea surface temperature. A band of warm waters snakes up the East Coast of the United States and veers across the North Atlantic the Gulf Stream.

Although short-lived weather events that influence ocean temperature are often washed out in monthly averages, a few events show up. For example, in December 2003, strong winds blew southwest from the Gulf of Mexico over Central America toward the Pacific Ocean, driving surface waters away from the coast, and allowing cold water from deeper in the ocean to well up to the surface. These winds are a recurring phenomenon in the area in the winter; they are known as Tehuano winds.

LOW PRESSURE

A low-pressure area, low, depression or cyclone is a region on the topographic map where the atmospheric pressure is lower than that of surrounding locations. Low-pressure systems form under areas of wind divergence that occur in the upper levels of the troposphere. The formation process of a low-pressure area is known as cyclogenesis. Within the field of meteorology, atmospheric divergence aloft occurs in two areas. The first area is on the east side of upper troughs, which form half of a Rossby wave within the Westerlies (a trough with large wavelength that extends through the troposphere). A second area of wind divergence aloft occurs ahead of embedded shortwave troughs, which are of smaller wavelength. Diverging winds aloft ahead of these troughs cause atmospheric lift within the troposphere below, which lowers surface pressures as upward motion partially counteracts the force of gravity.

A clockwise spinning low pressure area or cyclone off southern Australia. The center of the spiral-shaped cloud system is also the center of a low, and usually is where the pressure is lowest

Thermal lows form due to localized heating caused by greater sunshine over deserts and other land masses. Since localized areas of warm air are less dense than their surroundings, this warmer air rises, which lowers atmospheric pressure near that portion of the Earth's surface. Large-scale thermal lows over continents help drive monsoon circulations. Low-pressure areas can also form due to organized thunderstorm activity over warm water. When this occurs over the tropics in concert with the Intertropical Convergence Zone, it is known as a monsoon trough. Monsoon troughs reach their northerly extent in August and their southerly extent in February. When a convective low acquires a well-hot circulation in the tropics it is termed a tropical cyclone. Tropical cyclones can form during any month of the year globally, but can occur in either the northern or southern hemisphere during December.

This low-pressure system over Iceland spins counter-clockwise due to balance between the Coriolis force and the pressure gradient force

Atmospheric lift will also generally produce cloud cover through adiabatic cooling once the air becomes saturated as it rises, although the low-pressure area typically brings cloudy skies, which act to minimize diurnal temperature extremes. Since clouds reflect sunlight, incoming shortwave solar radiation decreases, which causes lower temperatures during the day. At night the absorptive effect of clouds on outgoing longwave radiation, such as heat energy from the surface, allows for warmer diurnal low temperatures in all seasons. The stronger the area of low pressure, the stronger the winds experienced in its vicinity. Globally, low-pressure systems are most frequently located over the Tibetan Plateau and in the lee of the Rocky mountains. In Europe (particularly in the British Isles and Netherlands), recurring low-pressure weather systems are typically known as "depressions".

Formation

Cyclogenesis is the development and strengthening of cyclonic circulations, or low-pressure areas, within the atmosphere. Cyclogenesis is the opposite of cyclolysis, and has an anticyclonic (high-pressure system) equivalent which deals with the formation of high-pressure areas—anticyclogenesis. Cyclogenesis is an umbrella term for several different processes, all of which result in the development of some sort of cyclone. Meteorologists use the term "cyclone" where circular pressure systems flow in the direction of the Earth's rotation, which normally coincides with areas of low pressure. The largest low-pressure systems are cold-core polar cyclones and extratropical cyclones which lie on the synoptic scale. Warm-core cyclones such as tropical cyclones, mesocyclones, and polar lows lie within the smaller mesoscale. Subtropical cyclones are of intermediate size. Cyclogenesis can occur at various scales, from the microscale to the synoptic scale. Larger-scale troughs, also called Rossby waves, are synoptic in scale. Shortwave troughs embedded within the flow around larger scale troughs are smaller in scale, or mesoscale in nature. Both Rossby waves and shortwaves embedded within the flow around Rossby waves migrate equatorward of the polar cyclones located in both the Northern and Southern hemispheres. All share one important aspect, that of upward vertical motion within the troposphere. Such upward motions decrease the mass of local atmospheric columns of air, which lowers surface pressure.

Extratropical cyclones form as waves along weather fronts due to a passing by shortwave aloft or upper level jet streak before occluding later in their life cycle as cold-core cyclones. Polar lows are small-scale, short-lived atmospheric low-pressure systems that occur over the ocean areas poleward of the main polar front in both the Northern and Southern Hemispheres. They are part of the larger class of mesoscale weather-systems. Polar lows can be difficult to detect using conventional weather reports and are a hazard to high-latitude operations, such as shipping and gas- and oil-platforms. They are vigorous systems that have near-surface winds of at least 17 metres per second (38 mph).

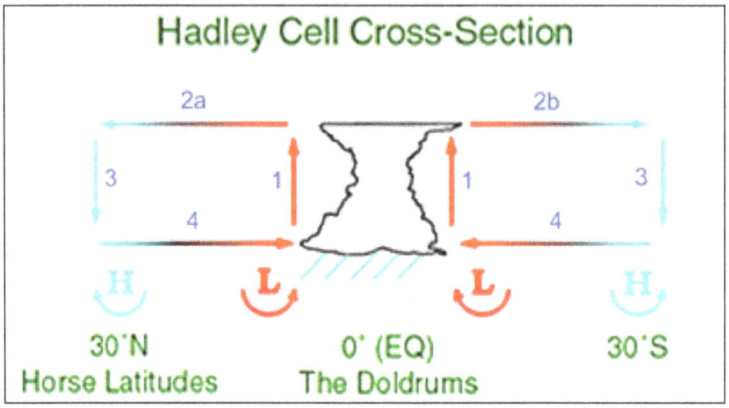

This depiction of the Hadley cell shows the process which sustains low-pressure areas. Diverging winds aloft allow for lower pressure and convergence at the Earth's surface, which leads to upward motion

Tropical cyclones form due to latent heat driven by significant thunderstorm activity, and are warm-core with well-defined circulations. Certain criteria need to be met for their formation. In most situations, water temperatures of at least 26.5 °C (79.7 °F) are needed down to a depth of at least 50 m (160 ft); waters of this temperature cause the overlying atmosphere to be unstable enough to sustain convection and thunderstorms. Another factor is rapid cooling with height, which allows the release of the heat of condensation that powers a tropical cyclone. High humidity is needed, especially in the lower-to-mid troposphere; when there is a great deal of moisture in the atmosphere, conditions are more favorable for disturbances to develop. Low amounts of wind shear are needed, as high shear is disruptive to the storm's circulation. Lastly, a formative tropical cyclone needs a pre-existing system of disturbed weather, although without a circulation no cyclonic development will take place. Mesocyclones form as warm core cyclones over land, and can lead to tornado formation. Waterspouts can also form from mesocyclones, but more often develop from environments of high instability and low vertical wind shear.

In deserts, lack of ground and plant moisture that would normally provide evaporative cooling can lead to intense, rapid solar heating of the lower layers of air. The hot air is less dense than surrounding cooler air. This, combined with the rising of the hot air, results in a low-pressure area called a thermal low. Monsoon circulations are caused by thermal lows which form over large areas of land and their strength is driven by how land heats more quickly than the surrounding nearby ocean. This generates a steady wind blowing toward the land, bringing the moist near-surface air over the oceans with it. Similar rainfall is caused by the moist ocean-air being lifted upwards by mountains, surface heating, convergence at the surface, divergence aloft, or from storm-produced outflows at the surface. However the lifting occurs, the air cools due to expansion in lower pressure, which in turn produces condensation. In winter, the land cools off quickly, but the ocean keeps the heat longer due to its higher specific heat. The hot air over the ocean rises, creating a low-pressure area and a breeze from land to ocean while a large area of drying high pressure is formed over the land, increased by wintertime cooling. Monsoons resemble sea and land breezes, terms usually referring to the localized, diurnal (daily) cycle of circulation near coastlines everywhere, but they are much larger in scale - also stronger and seasonal.

Climatology

Mid-latitudes and Subtropics

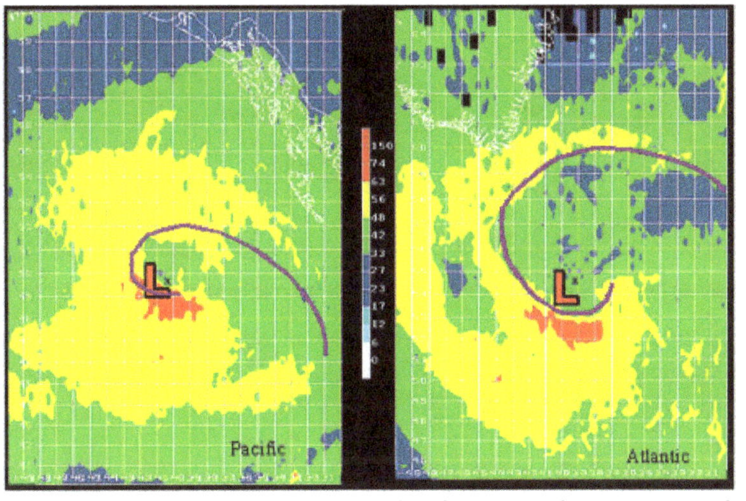

QuikSCAT image of typical extratropical cyclones over the ocean. Note the maximum winds on the poleward side of the occluded front

Large polar cyclones help determine the steering of systems moving through the mid-latitudes, south of the Arctic and north of the Antarctic. The Arctic oscillation provides an index used to gauge the magnitude of this effect in the Northern Hemisphere. Extratropical cyclones tend to form east of climatological trough positions aloft near the east coast of continents, or west side of oceans. A study of extratropical cyclones in the Southern Hemisphere shows that between the 30th and 70th parallels there are an average of 37 cyclones in existence during any 6-hour period. A separate study in the Northern Hemisphere suggests that approximately 234 significant extratropical cyclones form each winter. In Europe, particularly in the United Kingdom and in the Netherlands, recurring extratropical low-pressure weather systems are typically known as depressions. These tend to bring wet weather throughout the year. *Thermal lows* also occur during the summer over continental areas across the subtropics - such as the Sonoran Desert, the Mexican plateau, the Sahara, South America, and Southeast Asia. The lows are most commonly located over the Tibetan plateau and in the lee of the Rocky mountains.

Monsoon Trough

Elongated areas of low pressure form at the monsoon trough or intertropical convergence zone as part of the Hadley cell circulation. Monsoon troughing in the western Pacific reaches its zenith in latitude during the late summer when the wintertime surface ridge in the opposite hemisphere is the strongest. It can reach as far as the 40th parallel in East Asia during August and 20th parallel in Australia during February. Its poleward progression is accelerated by the onset of the summer monsoon which is characterized by the development of lower air pressure over the warmest part of the various continents. The large-scale thermal lows over continents help create pressure gradients which drive monsoon circulations. In the southern hemisphere, the monsoon trough associated with the Australian monsoon reaches its most southerly latitude in February, oriented along a west-northwest/east-southeast axis. Many of the world's rainforests are associated with these climatological low-pressure systems.

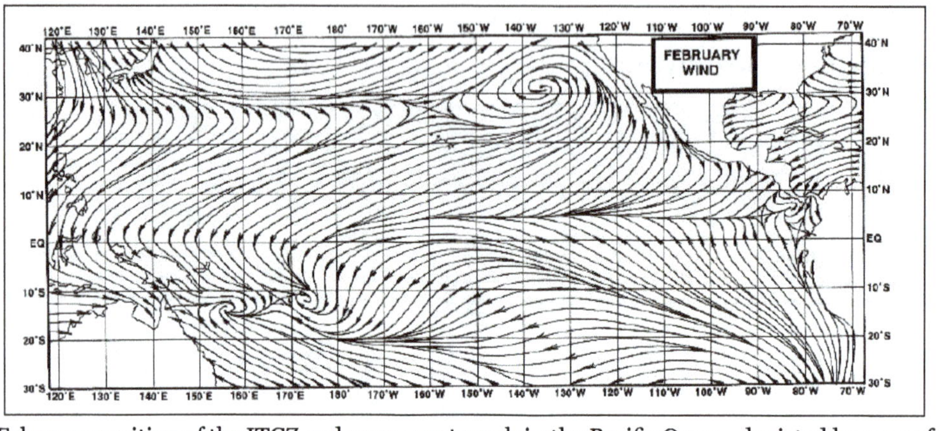

February position of the ITCZ and monsoon trough in the Pacific Ocean, depicted by area of convergent streamlines offshore Australia and in the equatorial eastern Pacific

Tropical Cyclone

Tropical cyclones generally need to form more than 555 km (345 mi) or poleward of the 5th parallel north and 5th parallel south, allowing the Coriolis effect to deflect winds blowing towards the low-pressure center and creating a circulation. Worldwide, tropical cyclone activity peaks in late summer, when the difference between temperatures aloft and sea surface temperatures is the greatest. However, each particular basin has its own seasonal patterns. On a worldwide scale, May is the least active month while September is the most active month. November is the only month that activity in all the tropical cyclone basins is possible. Nearly one-third of the world's tropical cyclones form within the western Pacific Ocean, making it the most active tropical cyclone basin on Earth.

Infrared image of a powerful southern hemisphere cyclone, Winston, near its initial peak intensity

Associated Weather

Wind is initially accelerated from areas of high pressure to areas of low pressure. This is due to density (or temperature and moisture) differences between two air masses. Since stronger high-pressure systems contain cooler or drier air, the air mass is denser and flows towards areas that are warm or moist, which are in the vicinity of low-pressure areas in advance of their associated cold fronts. The stronger the pressure difference, or pressure gradient, between a high-pressure system and a low-pressure system, the stronger the wind. Thus, stronger areas of low pressure are associated with stronger winds.

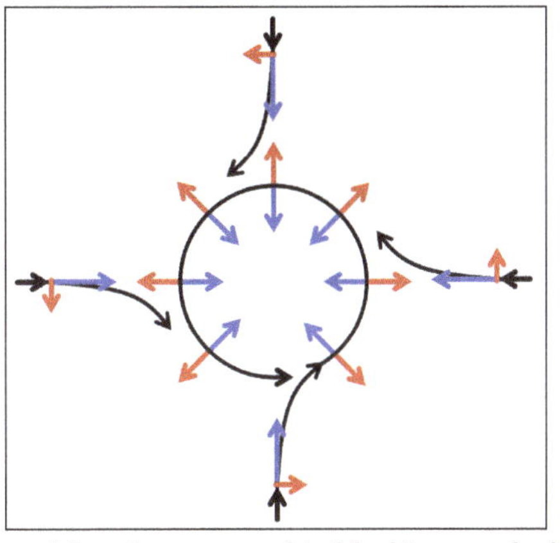

Schematic representation of flow (represented in black) around a low-pressure area in the Northern hemisphere. The pressure-gradient force is represented by blue arrows, the Coriolis acceleration (always perpendicular to the velocity) by red arrows

The Coriolis force caused by the Earth's rotation is what gives winds around low-pressure areas (such as in hurricanes, cyclones, and typhoons) their counter-clockwise (anticlockwise) circulation in the northern hemisphere (as the wind moves inward and is deflected right from the center of high pressure) and clockwise circulation in the southern hemisphere (as the wind moves inward and is deflected left from the center of high pressure). A cyclone differs from a hurricane or typhoon only on the basis of location. A hurricane is a storm that occurs in the Atlantic Ocean and northeastern Pacific Ocean, a typhoon occurs in the northwestern Pacific Ocean, and a cyclone occurs in the south Pacific or Indian Ocean. Friction with land slows down the wind flowing into low-pressure systems and causes wind to flow more inward, or flowing more ageostrophically, toward their centers. A low-pressure area is commonly associated with inclement weather, while a high-pressure area is associated with light winds and fair skies. Tornadoes are often too small, and of too short duration, to be influenced by the Coriolis force, but may be so-influenced when arising from a low-pressure system.

WEATHER FRONT

A weather front is a boundary separating two masses of air of different densities, and is the principal cause of meteorological phenomena outside the tropics. In surface weather analyses, fronts are depicted using various colored triangles and half-circles, depending on the type of front. The air masses separated by a front usually differ in temperature and humidity.

Cold fronts may feature narrow bands of thunderstorms and severe weather, and may on occasion be preceded by squall lines or dry lines. Warm fronts are usually preceded by stratiform precipitation and fog. The weather usually clears quickly after a front's passage. Some fronts produce no precipitation and little cloudiness, although there is invariably a wind shift.

Approaching weather fronts are often visible from the ground, but are not always as well defined as this

Cold fronts and occluded fronts generally move from west to east, while warm fronts move poleward. Because of the greater density of air in their wake, cold fronts and cold occlusions move faster than warm fronts and warm occlusions. Mountains and warm bodies of water can slow the movement of fronts. When a front becomes stationary—and the density contrast across the frontal boundary vanishes—the front can degenerate into a line which separates regions of differing wind velocity, known as a shearline. This is most common over the open ocean.

Bergeron Classification of Air Masses

The Bergeron classification is the most widely accepted form of air mass classification. Air mass classifications are indicated by three letters. The first letter describes its moisture properties, with c used for continental air masses (dry) and m for maritime air masses (moist). The second letter describes the thermal characteristic of its source region: T for tropical, P for polar, A for arctic or Antarctic, M for monsoon, E for equatorial, and S for superior air (dry air formed by significant upward motion in the atmosphere). The third letter designates the stability of the atmosphere. If the air mass is colder than the ground below it, it is labeled k. If the air mass is warmer than the ground below it, it is labeled w. Fronts separate air masses of different types or origins, and are located along troughs of lower pressure.

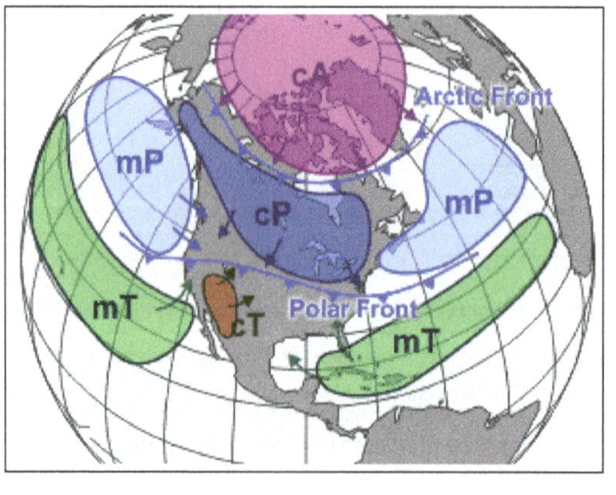

Different air masses which affect North America, as well as other continents, tend to be separated by frontal boundaries. In this, the Arctic front separates Arctic from Polar air masses, while the Polar front separates Polar air from warm air masses. (cA is continental arctic; cP is continental polar; mP is maritime polar; cT is continental tropic; and mT is maritime tropic.)

Surface Weather Analysis

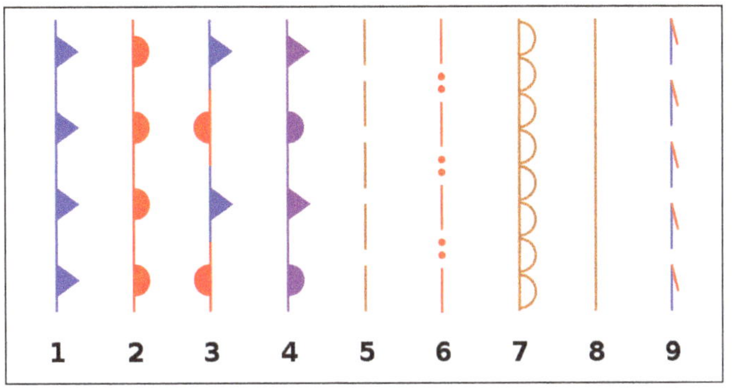

Weather map symbols:

1. Cold front;
2. Warm front;
3. Stationary front;
4. Occluded front;
5. Surface trough;
6. Squall/shear line;
7. Dry line;
8. Tropical wave;
9. Trowal.

A surface weather analysis is a special type of weather map which provides a view of weather elements over a geographical area at a specified time based on information from ground-based weather stations. Weather maps are created by plotting or tracing the values of relevant quantities such as sea-level pressure, temperature, and cloud cover onto a geographical map to help find synoptic scale features such as weather fronts. Surface weather analyses have special symbols which show frontal systems, cloud cover, precipitation, or other important information. For example, an *H* may represent high pressure, implying fair weather. An *L* on the other hand may represent low pressure, which frequently accompanies precipitation. Low pressure also creates surface winds deriving from high pressure zones. Various symbols are used not just for frontal zones and other surface boundaries on weather maps, but also to depict the present weather at various locations on the weather map. In addition, areas of precipitation help determine the frontal type and location.

Types of Weather Front

There are two different meanings used within meteorology to describe weather around a frontal zone. The term "anafront" describes boundaries which show instability, meaning air rises rapidly along and over the boundary to cause significant weather changes. A "katafront" is weaker, bringing smaller changes in temperature and moisture, as well as limited rainfall.

Cold Front

A cold front is located at the leading edge of the temperature drop off, which in an isotherm analysis shows up as the leading edge of the isotherm gradient, and it normally lies within a sharp surface trough. Cold fronts often bring heavy thunderstorms, rain, and hail. Cold fronts can produce sharper changes in weather and move up to twice as quickly as warm fronts, since cold air is denser than warm air and rapidly replaces the warm air preceding the boundary. On weather maps, the surface position of the cold front is marked with the symbol of a blue line of triangle-shaped pips pointing in the direction of travel, and it is placed at the leading edge of the cooler air mass. Cold fronts come in association with a low-pressure area. The concept of colder, dense air "wedging" under the less dense warmer air is often used to depict how air is lifted along a frontal boundary. The cold air wedging underneath warmer air creates the strongest winds just above the ground surface, a phenomenon often associated with property-damaging wind gusts. This lift would then form a narrow line of showers and thunderstorms if enough moisture were present. However, this concept isn't an accurate description of the physical processes; upward motion is not produced because of warm air "ramping up" cold, dense air, rather, frontogenetical circulation is behind the upward forcing.

Warm Front

Warm fronts are at the leading edge of a homogeneous warm air mass, which is located on the equatorward edge of the gradient in isotherms, and lie within broader troughs of low pressure than cold fronts. A warm front moves more slowly than the cold front which usually follows because cold air is denser and harder to remove from the Earth's surface.

This also forces temperature differences across warm fronts to be broader in scale. Clouds ahead of the warm front are mostly stratiform, and rainfall gradually increases as the front approaches. Fog can also occur preceding a warm frontal passage. Clearing and warming is usually rapid after frontal passage. If the warm air mass is unstable, thunderstorms may be embedded among the stratiform clouds ahead of the front, and after frontal passage thundershowers may continue. On weather maps, the surface location of a warm front is marked with a red line of semicircles pointing in the direction of travel.

Occluded Front

An occluded front is formed when a cold front overtakes a warm front, and usually forms around mature low-pressure areas. The cold and warm fronts curve naturally poleward into the point of occlusion, which is also known as the triple point. It lies within a sharp trough, but the air mass behind the boundary can be either warm or cold. In a cold occlusion, the air mass overtaking the warm front is cooler than the cool air ahead of the warm front and plows under both air masses. In a warm occlusion, the air mass overtaking the warm front is warmer than the cold air ahead of the warm front and rides over the colder air mass while lifting the warm air.

A wide variety of weather can be found along an occluded front, with thunderstorms possible, but usually their passage is associated with a drying of the air mass. Within the occlusion of the front, a circulation of air brings warm air upward and sends drafts of cold air downward, or vice versa depending on the occlusion the front is experiencing. Precipitations and clouds are associated with the *trowal*, the projection on the Earth's surface of the tongue of warm air aloft formed during the occlusion process of the depression.

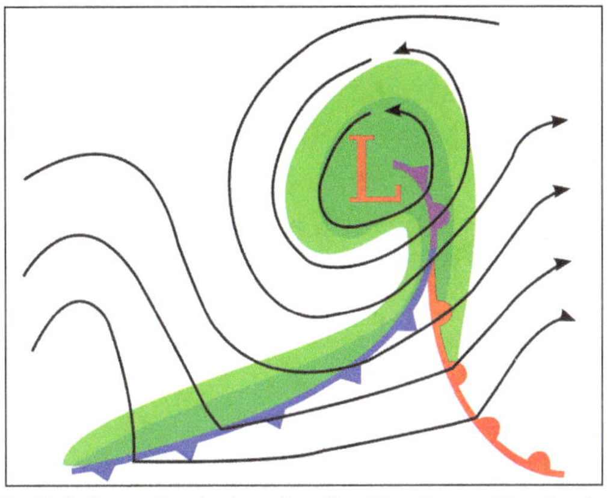

Occluded front depiction for the Northern Hemisphere

Occluded fronts are indicated on a weather map by a purple line with alternating half-circles and triangles pointing in direction of travel. The trowal is indicated by a series of blue and red junction lines.

Stationary Front

A stationary front is a non-moving (or stalled) boundary between two air masses, neither of which is strong enough to replace the other. They tend to remain essentially in the same area for extended periods of time, usually moving in waves. There is normally a broad temperature gradient behind the boundary with more widely spaced isotherm packing.

A wide variety of weather can be found along a stationary front, but usually clouds and prolonged precipitation are found there. Stationary fronts either dissipate after several days or devolve into shear lines, but they can transform into a cold or warm front if conditions aloft change. Stationary fronts are marked on weather maps with alternating red half-circles and blue spikes pointing in opposite directions, indicating no significant movement.

When stationary fronts become smaller in scale, degenerating to a narrow zone where wind direction changes significantly over a relatively short distance, they become known as shearlines. A shearline is depicted as a line of red dots and dashes. Stationary fronts may bring snow or rain for a long period of time.

Dry Line

A similar phenomenon to a weather front is the dry line, which is the boundary between air masses with significant moisture differences. When the westerlies increase on the north side of surface highs, areas of lowered pressure will form downwind of north–south oriented mountain chains, leading to the formation of a lee trough. Near the surface during daylight hours, warm moist air is denser than dry air of greater temperature, and thus the warm moist air wedges under the drier air like a cold front. At higher altitudes, the warm moist air is less dense than the dry air and the boundary slope reverses. In the vicinity of the reversal aloft, severe weather is possible, especially when a triple point is formed with a cold front. A weaker form of the dry line seen more commonly is the lee trough, which displays weaker differences in moisture. When moisture pools along the boundary during the warm season, it can be the focus of diurnal thunderstorms.

The dry line may occur anywhere on earth in regions intermediate between desert areas and warm seas. The southern plains west of the Mississippi River in the United States are a particularly favored location. The dry line normally moves eastward during the day and westward at night. A dry line is depicted on National Weather Service (NWS) surface analyses as an orange line with scallops facing into the moist sector. Dry lines are one of the few surface fronts where the pips indicated do not necessarily reflect the direction of motion.

Squall Line

Organized areas of thunderstorm activity not only reinforce pre-existing frontal zones, but can outrun cold fronts in a pattern where the upper level jet splits apart into two streams, with the resultant Mesoscale Convective System (MCS) forming at the point of the upper level split in the wind pattern running southeast into the warm sector parallel to low-level thickness lines. When the convection is strong and linear or curved, the MCS is called a squall line, with the feature placed at the leading edge of the significant wind shift and pressure rise. Even weaker and less organized areas of thunderstorms lead to locally cooler air and higher pressures, and outflow boundaries exist ahead of this type of activity, which can act as foci for additional thunderstorm activity later in the day.

A shelf cloud such as this one can be a sign that a squall is imminent

These features are often depicted in the warm season across the United States on surface analyses and lie within surface troughs. If outflow boundaries or squall lines form over arid regions, a haboob may result. Squall lines are depicted on NWS surface analyses as an alternating pattern of two red dots and a dash labelled SQLN or Squall Line, while outflow boundaries are depicted as troughs with a label of Outflow Boundary.

Precipitation Produced

Fronts are the principal cause of significant weather. *Convective precipitation* (showers, thundershowers, and related unstable weather) is caused by air being lifted and condensing into clouds by the movement of the cold front or cold occlusion under a mass of warmer, moist air. If the temperature differences of the two air masses involved are large and the turbulence is extreme because of wind shear and the presence of a strong jet stream, "roll clouds" and tornadoes may occur.

Convective precipitation

In the warm season, lee troughs, breezes, outflow boundaries and occlusions can lead to convection if enough moisture is available. *Orographic precipitation* is precipitation created through the lifting action of air moving over terrain such as mountains and hills, which is most common behind cold fronts that move into mountainous areas. It may sometimes occur in advance of warm fronts moving northward to the east of mountainous terrain. However, precipitation along warm fronts is relatively steady, as in rain or drizzle. Fog, sometimes extensive and dense, often occurs in pre-warm-frontal areas. Although, not all fronts produce precipitation or even clouds because moisture must be present in the air mass which is being lifted.

Movement

Fronts are generally guided by winds aloft, but do not move as quickly. Cold fronts and occluded fronts in the Northern Hemisphere usually travel from the northwest to southeast, while warm fronts move more poleward with time. In the Northern Hemisphere a warm front moves from southwest to northeast. In the Southern Hemisphere, the reverse is true; a cold front usually moves from southwest to northeast, and a warm front moves from northwest to southeast. Movement is largely caused by the pressure gradient force (horizontal differences in atmospheric pressure) and the Coriolis effect, which is caused by Earth's spinning about its axis. Frontal zones can be slowed down by geographic features like mountains and large bodies of warm water.

TROPICAL CYCLONE

Tropical Cyclones are low pressure systems that form over warm tropical waters and have gale force winds (sustained winds of 63 km/h or greater and gusts in excess of 90 km/h) near the centre. Technically they are defined as a non-frontal low pressure system of synoptic scale developing over warm waters having organised convection and a maximum mean wind speed of 34 knots or greater extending more than half-way around near the centre and persisting for at least six hours.

The gale force winds can extend hundreds of kilometres from the cyclone centre. If the sustained winds around the centre reach 118 km/h (gusts in excess 165 km/h) then the system is called a severe tropical cyclone. These are referred to as hurricanes or typhoons in other countries.

Structure of a cyclone

The circular eye or centre of a tropical cyclone is an area characterised by light winds and often by clear skies. Eye diameters are typically 40 km but can range from under 10 km to over 100 km. The eye is surrounded by a dense ring of cloud about 16 km high known as the eye wall which marks the belt of strongest winds and heaviest rainfall.

Tropical cyclones derive their energy from the warm tropical oceans and do not form unless the sea-surface temperature is above 26.5°C, although once formed, they can persist over lower sea-surface temperatures. Tropical cyclones can persist for many days and may follow quite erratic paths. They usually dissipate over land or colder oceans.

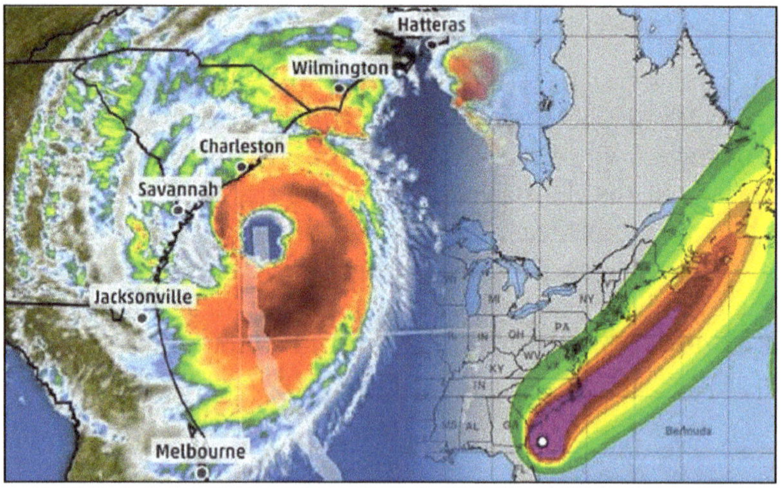

Radar image of cyclone

Cyclone Danger and Impacts

Tropical Cyclones are dangerous because they produce destructive winds, heavy rainfall with flooding and damaging storm surges that can cause inundation of low-lying coastal areas.

Cyclones have wind gusts in excess of 90 km/h around their centres and, in the most severe cyclones, gusts can exceed 280 km/h. These very destructive winds can cause extensive property damage and turn airborne debris into potentially lethal missiles. It is important to remember that, during the passage of the cyclone centre or *eye*, there will be a temporary lull in the wind, but that this will soon be replaced by destructive winds from another direction.

Heavy rainfall associated with the passage of a tropical cyclone can produce extensive flooding. This can cause further damage and death by drowning. The heavy rain can persist as the cyclone moves inland and decays, hence flooding due to a decayed cyclone can occur a long way from the tropical coast as the remains of a cyclone move into central and southern parts of the continent.

The destructive winds accompanying tropical cyclones also produce phenomenal seas, which are dangerous both for vessels out at sea and those moored in harbours. These seas can also cause serious erosion of foreshores.

Storm Surge and Tides

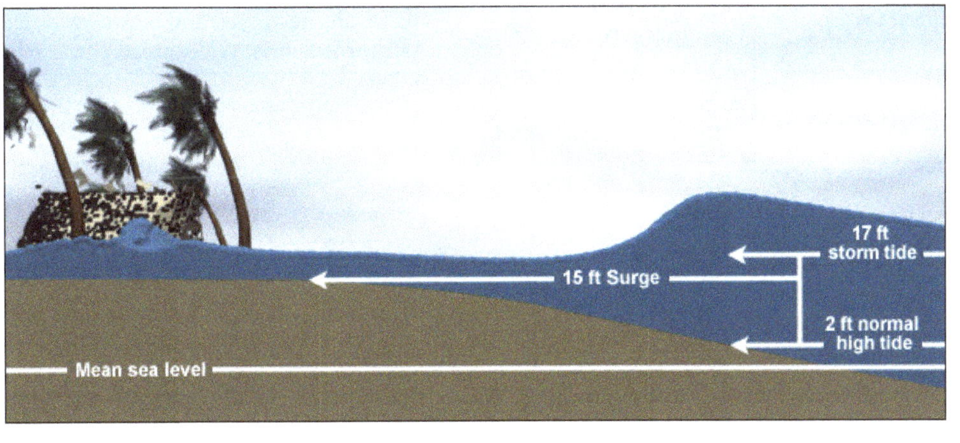

Potentially, the most destructive phenomenon associated with tropical cyclones that make landfall is the storm surge. Storm surge is a raised dome of water about 60 to 80 km across and typically about 2 to 5 m higher than the normal tide level. If the surge occurs at the same time as a high tide then the area inundated can be quite extensive, particularly along low-lying coastlines.

Tropical Cyclone Severity Categories

The severity of a tropical cyclone is described in terms of categories ranging from 1 to 5 related to the zone of maximum winds. An estimate of cyclone severity is included in all tropical advices. Remember that the Warning Service is not designed to give an exact statement of conditions at individual locations but will give a general idea of the expected worst conditions. Using this severity scale, communities will be able to assess the degree of cyclone threat and take appropriate action. Damage will vary depending upon factors such as:

- How far you are from the zone of maximum winds;
- How exposed the location is;
- Building standards;
- Vegetation type; and
- Resultant flooding.

The category does not refer to the amount of flooding or storm tides. If a storm tide is expected it will be mentioned separately in the cyclone warning.

Category	Strongest gust (km/h)	Typical effects
Tropical Cyclone	Less than 125 km/h Gales	Minimal house damage. Damage to some crops, trees and caravans.Boats may drag moorings.
Tropical Cyclone	125 - 164 km/h Destructive winds	Minor house damage. Significant damage to signs, trees and caravans. Heavy damage to some crops. Risk of power failure. Small boats may break moorings.
Severe Tropical Cyclone	165 - 224 km/h Very destructive winds	Some roof and structural damage. Some caravans destroyed. Power failure likely.
Severe Tropical Cyclone	225 - 279 km/h Very destructive winds	Significant roofing and structural damage. Many caravans destroyed and blown away. Dangerous airborne debris. Widespread power failures.
Severe Tropical Cyclone	More than 280 km/h Extremely destructive winds	Extremely dangerous with widespread destruction.

References

- Cloud, entry: newworldencyclopedia.org, Retrieved 25 June, 2019
- Feldman, D (February 25, 2015). "Observational determination of surface radiative forcing by CO2 from 2000 to 2010". Nature. 519 (7543): 339343. Bibcode:2015Natur.519..339F. Doi:10.1038/nature14240. PMID 25731165
- Freezing-rain, rain, types-of-weather, weather, learn-about, weather, metoffice.gov.uk, Retrieved 26 July, 2019
- Steve Kempler (2009). "Parameter information page". NASA Goddard Space Flight Center. Archived from the original on November 26, 2007. Retrieved 2008-12-27
- Freezing-Rain: skybrary.aero, Retrieved 28 August, 2019
- Scott, A (2000). "The Pre-Quaternary history of fire". Palaeogeography, Palaeoclimatology, Palaeoecology. 164 (1–4): 281–329. Bibcode:2000PPP...164..281S. Doi:10.1016/S0031-0182(00)00192-9

- Sea-surface-temperature, measurements: measurements, Retrieved 27 July, 2019
- Arctic Climatology and Meteorology (2009). Cyclogenesis.Archived 2006-08-30 at the Wayback Machine National Snow and Ice Data Center. Retrieved on 2009-02-21
- Global-maps: earthobservatory.nasa.gov, Retrieved 29 January, 2019
- Jon Nelson (September 26, 2008). "Origin of diversity in falling snow". Atmospheric Chemistry and Physics. 8 (18): 5669–5682. Doi:10.5194/acp-8-5669-2008
- About, cyclone: bom.gov.au, Retrieved 30 March, 2019
- Glossary of Meteorology (2009). "Geostrophic wind". American Meteorological Society. Archived from the originalon 2007-10-16. Retrieved 2009-03-18

Branches of Meteorology

CHAPTER 3

Some of the major branches of meteorology are microscale meteorology, mesoscale meteorology, synoptic scale meteorology and hydrometeorology. The chapter closely examines the key concepts of these branches of meteorology to provide an extensive understanding of the subject.

MICROSCALE METEOROLOGY

Microscale meteorology is the study of short-lived atmospheric phenomena smaller than mesoscale, about 1 km or less. These two branches of meteorology are sometimes grouped together as "mesoscale and microscale meteorology" (MMM) and together study all phenomena smaller than synoptic scale; that is they study features generally too small to be depicted on a weather map. These include small and generally fleeting cloud "puffs" and other small cloud features. Microscale meteorology controls the most important mixing and dilution processes in the atmosphere. Important topics in microscale meteorlogy include heat transfer and gas exchange between soil, vegetation, and/or surface water and the atmosphere caused by near-ground turbulence. Measuring these transport processes involves use of micrometeorological (or flux) towers. Variables often measured or derived include net radiation, sensible heat flux, latent heat flux, ground heat storage, and fluxes of trace gases important to the atmosphere, biosphere, and hydrosphere.

Meteorological analysis and prediction of microscale weather events like tornados, microburst, and other localized effects can be used as a predictive tool, or retrospectively to evaluate historical events. Applications include asset management, corporate risk management, strategic planning, building and engineering design assessment, and determination of agricultural impacts, to name a few.

MESOSCALE METEOROLOGY

Mesoscale meteorology is the study of weather systems smaller than synoptic scale systems but larger than microscale and storm-scale cumulus systems. Horizontal dimensions generally range from around 5 kilometers to several hundred kilometers. Examples of mesoscale weather systems are sea breezes, squall lines, and mesoscale convective complexes.

Vertical velocity often equals or exceeds horizontal velocities in mesoscale meteorological systems due to nonhydrostatic processes such as buoyant acceleration of a rising thermal or acceleration through a narrow mountain pass.

A meso-beta scale vortex

Subclasses

Mesoscale Meteorology is divided into these subclasses:

- Meso-alpha 200–2000 km scale of phenomena like fronts, squall lines, mesoscale convective systems (MCS), tropical cyclones at the edge of synoptic scale.
- Meso-beta 20–200 km scale of phenomena like sea breezes, lake effect snow storms.
- Meso-gamma 2–20 km scale of phenomena like thunderstorm convection, complex terrain flows (at the edge to microscale, also known as storm-scale).

As a note, tropical and subtropical cyclones are classified by National Hurricane Center as synoptic scale rather than mesoscale.

Mesoscale Boundaries

As in synoptic frontal analysis, mesoscale analysis uses cold, warm, and occluded fronts on the mesoscale to help describe phenomena. On weather maps mesoscale fronts are depicted as smaller and with twice as many bumps or spikes as the synoptic variety. In the United States, opposition to the use of the mesoscale versions of fronts on weather analyses, has led to the use of an overarching symbol (a trough symbol) with a label of outflow boundary as the frontal notation.

SYNOPTIC SCALE METEOROLOGY

The synoptic scale in meteorology (also known as large scale or cyclonic scale) is a horizontal length scale of the order of 1000 kilometers (about 620 miles) or more. This corresponds to a horizontal scale typical of mid-latitude depressions (e.g., extratropical cyclones). Most high and low-pressure areas seen on weather maps such as surface weather analyses are synoptic-scale systems, driven by the location of Rossby waves in their respective hemisphere. Low-pressure areas and their related frontal zones occur on the leading edge of a trough within the Rossby wave pattern, while high-pressure areas form on the back edge of the trough. Most precipitation areas occur near frontal zones. The word *synoptic* is derived from the Greek word *synoptikos*, meaning *seen together*.

The Navier–Stokes equations applied to atmospheric motion can be simplified by scale analysis in the synoptic scale. It can be shown that the main terms in horizontal equations are Coriolis force and pressure gradient terms; therefore, one can use geostrophic approximation. In vertical coordinates, the momentum equation simplifies to the hydrostatic equilibrium equation.

Surface Weather Analysis

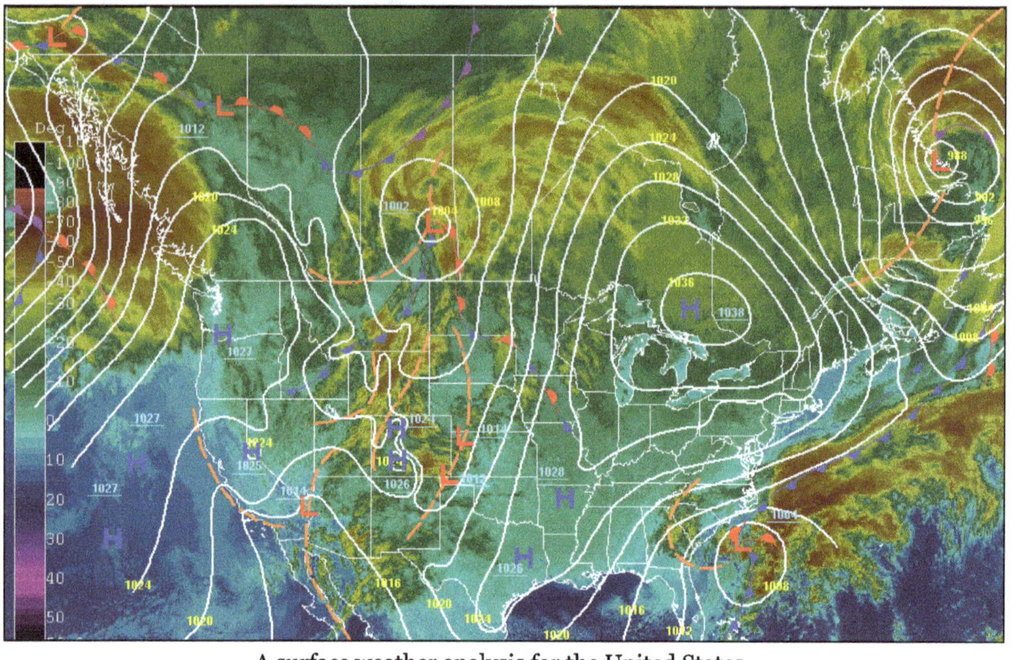

A surface weather analysis for the United States

A surface weather analysis is a special type of weather map that provides a view of weather elements over a geographical area at a specified time based on information from ground-based weather stations. Weather maps are created by plotting or tracing the values of relevant quantities such as sea level pressure, temperature, and cloud cover onto a geographical map to help find synoptic scale features such as weather fronts.

The first weather maps in the 19th century were drawn well after the fact to help devise a theory on storm systems. After the advent of the telegraph, simultaneous surface weather observations became possible for the first time, and beginning in the late 1840s, the Smithsonian Institution became the first organization to draw real-time surface analyses. Use of surface analyses began first in the United States, spreading worldwide during the 1870s. Use of the Norwegian cyclone model for frontal analysis began in the late 1910s across Europe, with its use finally spreading to the United States during World War II.

Surface weather analyses have special symbols which show frontal systems, cloud cover, precipitation, or other important information. For example, an *H* represents high pressure, implying good and fair weather. An *L* represents low pressure, which frequently accompanies precipitation. Various symbols are used not just for frontal zones and other surface boundaries on weather maps, but also to depict the present weather at various locations on the weather map. Areas of precipitation help determine the frontal type and location. Mesoscale systems and boundaries such as tropical cyclones, outflow boundaries and squall lines also are analyzed on surface weather analyses.

Isobars are commonly used to place surface boundaries from the horse latitudes poleward, while streamline analyses are used in the tropics.

Extratropical Cyclone

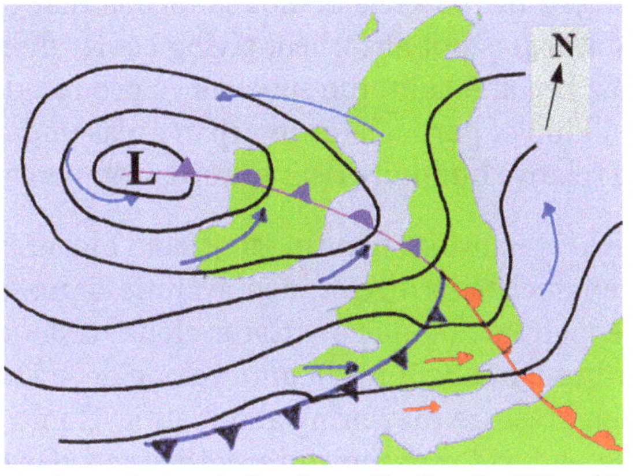

A fictitious synoptic chart of an extratropical cyclone affecting the UK and Ireland. The blue arrows between isobars indicate the direction of the wind, while the "L" symbol denotes the centre of the "low". Note the occluded, cold and warm frontal boundaries

An extratropical cyclone is a synoptic scale low-pressure weather system that has neither tropical nor polar characteristics, being connected with fronts and horizontal gradients in temperature and dew point otherwise known as "baroclinic zones".

The descriptor "extratropical" refers to the fact that this type of cyclone generally occurs outside of the tropics, in the middle latitudes of the planet. These systems may also be described as "mid-latitude cyclones" due to their area of formation, or "post-tropical cyclones" where extratropical transition has occurred, and are often described as "depressions" or "lows" by weather forecasters and the public. These are the everyday phenomena which along with anti-cyclones, drive the weather over much of the Earth.

Although extratropical cyclones are almost always classified as baroclinic since they form along zones of temperature and dewpoint gradient within the westerlies, they can sometimes become barotropic late in their life cycle when the temperature distribution around the cyclone becomes fairly uniform with radius. An extratropical cyclone can transform into a subtropical storm, and from there into a tropical cyclone, if it dwells over warm waters and develops central convection, which warms its core.

Surface High-pressure Systems

Golden Gate Bridge in fog

High-pressure systems are frequently associated with light winds at the surface and subsidence through the lower portion of the troposphere. Subsidence will generally dry out an air mass by adiabatic, or compressional, heating. Thus, high pressure typically brings clear skies. During the day, since no clouds are present to reflect sunlight, there is more incoming shortwave solar radiation and temperatures rise. At night, the absence of clouds means that outgoing longwave radiation (i.e. heat energy from the surface) is not absorbed, giving cooler diurnal low temperatures in all seasons. When surface winds become light, the subsidence produced directly under a high-pressure system can lead to a buildup of particulates in urban areas under the ridge, leading to widespread haze. If the low level relative humidity rises towards 100 percent overnight, fog can form.

Strong, vertically shallow high-pressure systems moving from higher latitudes to lower latitudes in the northern hemisphere are associated with continental arctic air masses. The low, sharp inversion can lead to areas of persistent stratocumulus or stratus cloud, colloquially known as anticyclonic gloom. The type of weather brought about by an anticyclone depends on its origin. For example, extensions of the Azores high pressure may bring about anticyclonic gloom during the winter, as they are warmed at the base and will trap moisture as they move over the warmer oceans. High pressures that build to the north and extend southwards will often bring clear weather. This is due to being cooled at the base (as opposed to warmed) which helps prevent clouds from forming.

On weather maps, these areas show converging winds (isotachs), also known as confluence, or converging height lines near or above the level of non-divergence, which is near the 500 hPa pressure surface about midway up through the troposphere. High-pressure systems are alternatively referred to as anticyclones. On weather maps, high-pressure centers are associated with the letter H in English, or A in Spanish, because alta is the Spanish word for high, within the isobar with the highest pressure value. On constant pressure upper level charts, it is located within the highest height line contour.

Weather Fronts

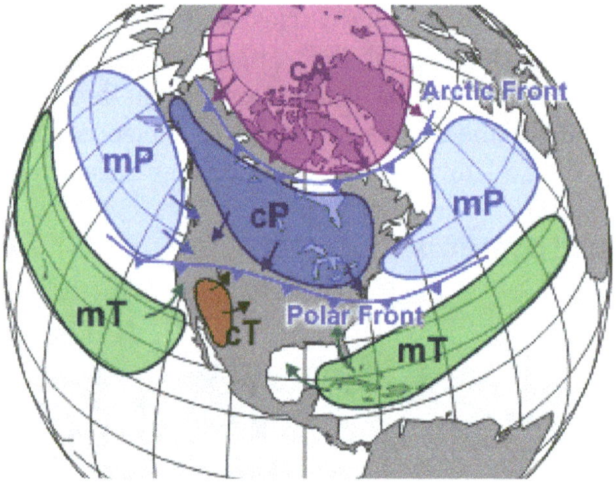

Different air masses tend to be separated by frontal boundaries. The Arctic front separates Arctic from Polar air masses, while the Polar front separates Polar air from warm air masses. (cA is continental arctic; cP is continental polar; mP is maritime polar; cT is continental tropic; and mT is maritime tropic.)

A weather front is a boundary separating two masses of air of different densities, and is the principal cause of meteorological phenomena. In surface weather analyses, fronts are depicted using various colored lines and symbols, depending on the type of front. The air masses separated by a front usually differ in temperature and humidity. Cold fronts may feature narrow bands of thunderstorms and severe weather, and may on occasion be preceded by squall lines or dry lines. Warm fronts are usually preceded by stratiform precipitation and fog. The weather usually clears quickly after a front's passage. Some fronts produce no precipitation and little cloudiness, although there is invariably a wind shift.

Cold fronts and occluded fronts generally move from west to east, while warm fronts move poleward. Because of the greater density of air in their wake, cold fronts and cold occlusions move faster than warm fronts and warm occlusions. Mountains and warm bodies of water can slow the movement of fronts. When a front becomes stationary, and the density contrast across the frontal boundary vanishes, the front can degenerate into a line which separates regions of differing wind velocity, known as a shearline. This is most common over the open ocean.

HYDROMETEOROLOGY

Hydrometeorology is the branch of meteorology that deals with problems involving the hydrologic cycle, the water budget, and the rainfall statistics of storms. The boundaries of hydrometeorology are not clear-cut, and the problems of the hydrometeorologist overlap with those of the climatologist, the hydrologist, the cloud physicist, and the weather forecaster. Considerable emphasis is placed on determining, theoretically or empirically, the relationships between meteorological variables and the maximum precipitation reaching the ground. These analyses often serve as the bases for the design of flood-control and water-usage structures, primarily dams and reservoirs. Other concerns of hydrometeorologists include the determination of rainfall probabilities, the space and time distribution of rainfall and evaporation, the recurrence interval of major storms, snow melt and runoff, and probable wind tides and waves in reservoirs. The whole field of water quality and supply is of growing importance in hydrometeorology.

References

- Atlantic Oceanographic and Meteorological Laboratory, Hurricane Research Division. "Frequently Asked Questions: What is an extra-tropical cyclone?". NOAA. Retrieved 2006-07-25
- Microscale-meteorology, service: novusenv.com, Retrieved 1 April, 2019
- Roth, David. "Unified Surface Analysis Manual" (PDF). Hydrometeorological Prediction Center. Retrieved 2006-10-24
- Hydrometeorology, science: britannica.com, Retrieved 2 June, 2019
- Dr. Decaria (2005-12-07). "ESCI 241 – Meteorology; Lesson 16 – Extratropical Cyclones". Department of Earth Sciences, Millersville University, Millersville, Pennsylvania. Archived from the original on 2006-09-03. Retrieved 2006-10-21

Weather Forecasting

CHAPTER 4

The application of science and technology for the prediction of atmospheric conditions of a given location and time is referred to as weather forecasting. Various types of weather forecasting include marine weather forecasting, numerical weather prediction and nowcasting. The diverse applications of these types of weather forecasting have been thoroughly discussed in this chapter.

Weather forecasting is the prediction of the weather through application of the principles of physics, supplemented by a variety of statistical and empirical techniques. In addition to predictions of atmospheric phenomena themselves, weather forecasting includes predictions of changes on Earth's surface caused by atmospheric conditions—e.g., snow and ice cover, storm tides, and floods.

Measurements and Ideas as the basis for Weather Prediction

The observations of few other scientific enterprises are as vital or affect as many people as those related to weather forecasting. From the days when early humans ventured from caves and other natural shelters, perceptive individuals in all likelihood became leaders by being able to detect nature's signs of impending snow, rain, or wind, indeed of any change in weather. With such information they must have enjoyed greater success in the search for food and safety, the major objectives of that time.

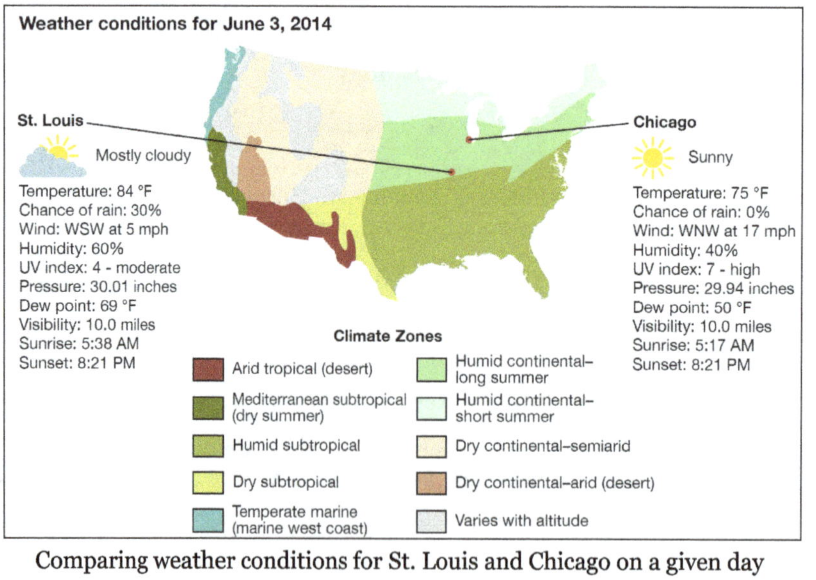

Comparing weather conditions for St. Louis and Chicago on a given day

In a sense, weather forecasting is still carried out in basically the same way as it was by the earliest humans—namely, by making observations and predicting changes. The modern tools used to measure temperature, pressure, wind, and humidity in the 21st century would certainly amaze them, and the results obviously are better. Yet, even the most sophisticated numerically calculated forecast made on a supercomputer requires a set of measurements of the condition of the atmosphere—an initial picture of temperature, wind, and other basic elements, somewhat comparable to that formed by our forebears when they looked out of their cave dwellings. The primeval approach entailed insights based on the accumulated experience of the perceptive observer, while the modern technique consists of solving equations. Although seemingly quite different, there are underlying similarities between both practices. In each case the forecaster asks "What is?" in the sense of "What kind of weather prevails today?" and then seeks to determine how it will change in order to extrapolate what it will be.

Because observations are so critical to weather prediction, an account of meteorological measurements and weather forecasting is a story in which ideas and technology are closely intertwined, with creative thinkers drawing new insights from available observations and pointing to the need for new or better measurements, and technology providing the means for making new observations and for processing the data derived from measurements. The basis for weather prediction started with the theories of the ancient Greek philosophers and continued with Renaissance scientists, the scientific revolution of the 17th and 18th centuries, and the theoretical models of 20th- and 21st-century atmospheric scientists and meteorologists. Likewise, it tells of the development of the "synoptic" idea—that of characterizing the weather over a large region at exactly the same time in order to organize information about prevailing conditions. In synoptic meteorology, simultaneous observations for a specific time are plotted on a map for a broad area whereby a general view of the weather in that region is gained. (The term synoptic is derived from the Greek word meaning "general or comprehensive view.") The so-called synoptic weather map came to be the principal tool of 19th-century meteorologists and continues to be used today in weather stations and on television weather reports around the world.

Since the mid-20th century, digital computers have made it possible to calculate changes in atmospheric conditions mathematically and objectively—i.e., in such a way that anyone can obtain the same result from the same initial conditions. The widespread adoption of numerical weather prediction models brought a whole new group of players—computer specialists and experts in numerical processing and statistics—to the scene to work with atmospheric scientists and meteorologists. Moreover, the enhanced capability to process and analyze weather data stimulated the long-standing interest of meteorologists in securing more observations of greater accuracy. Technological advances since the 1960s led to a growing reliance on remote sensing, particularly the gathering of data with specially instrumented Earth-orbiting satellites. By the late 1980s, forecasts of the weather were largely based on the determinations of numerical models integrated by high-speed supercomputers—except for some shorter-range predictions, particularly those related to local thunderstorm activity, which were made by specialists directly interpreting radar and satellite measurements. By the early 1990s a network of next-generation Doppler weather radar (NEXRAD) was largely in place in the United States, which allowed meteorologists to predict severe weather events with additional lead time before their occurrence. During the late 1990s and early 21st century, computer processing power increased, which allowed weather bureaus to produce more-sophisticated ensemble forecasts—that is, sets of multiple model runs whose results limit the range of uncertainty with respect to a forecast.

Practical Applications

Systematic weather records were kept after instruments for measuring atmospheric conditions became available during the 17th century. Undoubtedly these early records were employed mainly by those engaged in agriculture. Planting and harvesting obviously can be planned better and carried out more efficiently if long-term weather patterns can be estimated. In the United States, the foundations of the national weather services were laid down by American physicist Joseph Henry, the first head of the Smithsonian Institution. In 1849 Henry created a network of volunteer weather observers to help improve storm prediction in the U.S. The first national weather services were provided by the U.S. Army Signal Corps beginning on February 9, 1870, which also incorporated Henry's volunteer weather observers by 1874. These operations were taken over by the Department of Agriculture in 1891. By the early 1900s free mail service and telephone were providing forecasts daily to millions of American farmers. The U.S. Weather Bureau established a Fruit-Frost (forecasting) Service during World War I, and by the 1920s radio broadcasts to agricultural interests were being made in most states.

Weather forecasting became an important tool for aviation during the 1920s and '30s. Its application in this area gained in importance after Francis W. Reichelderfer was appointed chief of the U.S. Weather Bureau (USWB) in 1939. Reichelderfer had previously modernized the U.S. Navy's meteorological service and made it a model of support for naval aviation. During World War II the discovery of very strong wind currents at high altitudes (the jet streams, which can affect aircraft speed) and the general susceptibility of military operations in Europe to weather led to a special interest in weather forecasting.

One of the most famous wartime forecasting problems was for Operation Overlord, the invasion of the European mainland at Normandy by Allied forces. An unusually intense June storm brought high seas and gales to the French coast, but a moderation of the weather that was successfully predicted by Col. J.M. Stagg of the British forces (after consultation with both British and American forecasters) enabled Gen. Dwight D. Eisenhower, supreme commander of the Allied Expeditionary Forces, to make his critical decision to invade on June 6, 1944.

The second half of the 20th century saw a reorganization of the country's weather bureau. The USWB was part of the Department of Agriculture until 1940, when it was added to the Department of Commerce. On October 9, 1970, the USWB became the National Weather Service.

In addition, the later part of the 20th century was a time of unprecedented growth of commercial weather-forecasting firms in the United States and elsewhere. Marketing organizations and stores hire weather-forecasting consultants to help with the timing of sales and promotions of products ranging from snow tires and roofing materials to summer clothes and resort vacations. Many oceangoing shipping vessels as well as military ships use optimum ship routing forecasts to plan their routes in order to minimize lost time, potential damage, and fuel consumption in heavy seas. Similarly, airlines carefully consider atmospheric conditions when planning long-distance flights so as to avoid the strongest head winds and to ride with the strongest tail winds.

International trading of foodstuffs such as wheat, corn (maize), beans, sugar, cocoa, and coffee can be severely affected by weather news. For example, in 1975 a severe freeze in Brazil caused the price of coffee to increase substantially within just a few weeks, and in 2017 Georgia peach growers blamed the combination of warm winter temperatures and a spring freeze on the loss of nearly 80 percent of the state's peach crop. In addition, extreme heat and drought can affect production; one study estimated that 9–10 percent of cereal crops between 1964 and 2007 were lost to these phenomena. Weather-forecasting organizations are thus frequently called upon by banks, commodity traders, and food companies to give them advance knowledge of the possibility of such sudden changes. The cost of all sorts of commodities and services, whether they are tents for outdoor events or plastic covers for the daily newspapers, can be reduced or eliminated if reliable information about possible precipitation can be obtained in advance.

Forecasts must be quite precise for applications that are tailored to specific industries. Gas and electric utilities, for example, may require forecasts of temperature within one or two degrees a day ahead of time, or ski-resort operators may need predictions of nighttime relative humidity on the slopes within 5 to 10 percent in order to schedule snow making.

Emergence of Synoptic Forecasting Methods

Analysis of Synoptic Weather Reports

An observant person who has learned nature's signs can interpret the appearance of the sky, the wind, and other local effects and "foretell the weather." A scientist can use instruments at one location to do so even more effectively. The modern approach to weather forecasting, however, can only be realized when many such observations are exchanged quickly by experts at various weather stations and entered on a synoptic weather map to depict the patterns of pressure, wind, temperature, clouds, and precipitation at a specific time. Such a rapid exchange of weather data became feasible with the development of the electric telegraph in 1837 by Samuel F.B. Morse of the United States. By 1849 Joseph Henry of the Smithsonian Institution in Washington, D.C., was plotting daily weather maps based on telegraphic reports, and in 1869 Cleveland Abbe at the Cincinnati Observatory began to provide regular weather forecasts using data received telegraphically.

Synoptic weather maps resolved one of the great controversies of meteorology—namely, the rotary storm dispute. By the early decades of the 19th century, it was known that storms were associated with low barometric readings, but the relation of the winds to low-pressure systems, called cyclones, remained unrecognized. William Redfield, a self-taught meteorologist from Middletown, Conn., noticed the pattern of fallen trees after a New England hurricane and suggested in 1831 that the wind flow was a rotary counterclockwise circulation around the centre of lowest pressure. The American meteorologist James P. Espy subsequently proposed in his Philosophy of Storms that air would flow toward the regions of lowest pressure and then would be forced upward, causing clouds and precipitation. Both Redfield and Espy proved to be right. The air does spin around the cyclone, as Redfield believed, while the layers close to the ground flow inward and upward as well. The net result is a rotational wind circulation that is slightly modified at Earth's surface to produce inflow toward the storm centre, just as Espy had proposed. Further, the inflow is associated with clouds and precipitation in regions of low pressure, though that is not the only cause of clouds there.

In Europe the writings of Heinrich Dove, a Polish scientist who directed the Prussian Meteorological Institute, greatly influenced views concerning wind behaviour in storms. Unlike the Americans, Dove did not focus on the pattern of the winds around the storm but rather on how the wind should change at one place as a storm passed. It was many years before his followers understood the complexity of the possible changes.

Establishment of Weather-station Networks and Services

Routine production of synoptic weather maps became possible after networks of stations were organized to take measurements and report them to some type of central observatory. As early as 1814, U.S. Army Medical Corps personnel were ordered to record weather data at their posts; this activity was subsequently expanded and made more systematic. Actual weather-station networks were established in the United States by New York University, the Franklin Institute, and the Smithsonian Institution during the early decades of the 19th century.

In Britain, James Glaisher organized a similar network, as did Christophorus H.D. Buys Ballot in the Netherlands. Other such networks of weather stations were developed near Vienna, Paris, and St. Petersburg.

It was not long before national meteorological services were established on the Continent and in the United Kingdom. The first national weather service in the United States commenced operations in 1871, with responsibility assigned to the U.S. Army Signal Corps. The original purpose of the service was to provide storm warnings for the Atlantic and Gulf coasts and for the Great Lakes. Within the next few decades, national meteorological services were established in such countries as Japan, India, and Brazil. The importance of international cooperation in weather prognostication was recognized by the directors of such national services. By 1880 they had formed the International Meteorological Organization (IMO).

The proliferation of weather-station networks linked by telegraphy made synoptic forecasting a reality by the close of the 19th century. Yet, the daily weather forecasts generated left much to be desired. Many errors occurred as predictions were largely based on the experience that each individual forecaster had accumulated over several years of practice, vaguely formulated rules of thumb (e.g., of how pressure systems move from one region to another), and associations that were poorly understood, if at all.

Progress during the Early 20th Century

An important aspect of weather prediction is to calculate the atmospheric pressure pattern—the positions of the highs and lows and their changes. Modern research has shown that sea-level pressure patterns respond to the motions of the upper-atmospheric winds, with their narrow, fast-moving jet streams and waves that propagate through the air and pass air through themselves.

Frequent surprises and errors in estimating surface atmospheric pressure patterns undoubtedly caused 19th-century forecasters to seek information about the upper atmosphere for possible explanations. The British meteorologist Glaisher made a series of ascents by balloon during the 1860s, reaching an unprecedented height of nine kilometres. At about this time investigators on the Continent began using unmanned balloons to carry recording barographs, thermographs, and hygrographs to high altitudes. During the late 1890s meteorologists in both the United States and Europe used kites equipped with instruments to probe the atmosphere up to altitudes of about three kilometres. Notwithstanding these efforts, knowledge about the upper atmosphere remained very limited at the turn of the century. The situation was aggravated by the confusion created by observations from weather stations located on mountains or hilltops. Such observations often did not show what was expected, partly because so little was known about the upper atmosphere and partly because the mountains themselves affect measurements, producing results that are not representative of what would be found in the free atmosphere at the same altitude.

Fortunately, a large enough number of scientists had already put forth ideas that would make it possible for weather forecasters to think three-dimensionally, even if sufficient meteorological measurements were lacking. Henrik Mohn, the first of a long line of highly creative Norwegian meteorologists, Wladimir Köppen, the noted German climatologist, and Max Margules, an influential Russian-born meteorologist, all contributed to the view that mechanisms of the upper air generate the energy of storms.

In 1911 William H. Dines, a British meteorologist, published data that showed how the upper atmosphere compensates for the fact that the low-level winds carry air toward low-pressure centres. Dines recognized that the inflow near the ground is more or less balanced by a circulation upward and outward aloft. Indeed, for a cyclone to intensify, which would require a lowering of central pressure, the outflow must exceed the inflow; the surface winds can converge quite strongly toward the cyclone, but sufficient outflow aloft can produce falling pressure at the centre.

Meteorologists of the time were now aware that vertical circulations and upper-air phenomena were important, but they still had not determined how such knowledge could improve weather forecasting. Then, in 1919, the Norwegian meteorologist Jacob Bjerknes introduced what has been referred to as the Norwegian cyclone model. This theory pulled together many earlier ideas and related the patterns of wind and weather to a low-pressure system that exhibited fronts—which are rather sharp sloping boundaries between cold and warm air masses. Bjerknes pointed out the rainfall/snowfall patterns that are characteristically associated with the fronts in cyclones: the rain or snow occurs over large areas on the cold side of an advancing warm front poleward of a low-pressure centre. Here, the winds are from the lower latitudes, and the warm air, being light, glides up over a large region of cold air. Widespread, sloping clouds spread ahead of the cyclone; barometers fall as the storm approaches, and precipitation from the rising warm air falls through the cold air below. Where the cold air advances to the rear of the storm, squalls and showers mark the abrupt lifting of the warm air being displaced. Thus, the concept of fronts focused attention on the action at air mass boundaries. The Norwegian cyclone model could be called the frontal model, for the idea of warm air masses being lifted over cold air along their edges (fronts) became a major forecasting tool. The model not only emphasized the idea but it also showed how and where to apply it.

In later work, Bjerknes and several other members of the so-called Bergen school of meteorology expanded the model to show that cyclones grow from weak disturbances on fronts, pass through a regular life cycle, and ultimately die by the inflow filling them. Both the Norwegian cyclone model and the associated life-cycle concept are still used today by weather forecasters.

While Bjerknes and his Bergen colleagues refined the cyclone model, other Scandinavian meteorologists provided much of the theoretical basis for modern weather prediction. Foremost among them were Vilhelm Bjerknes, Jacob's father, and Carl-Gustaf Rossby. Their ideas helped make it possible to understand and carefully calculate the changes in atmospheric circulation and the motion of the upper-air waves that control the behaviour of cyclones.

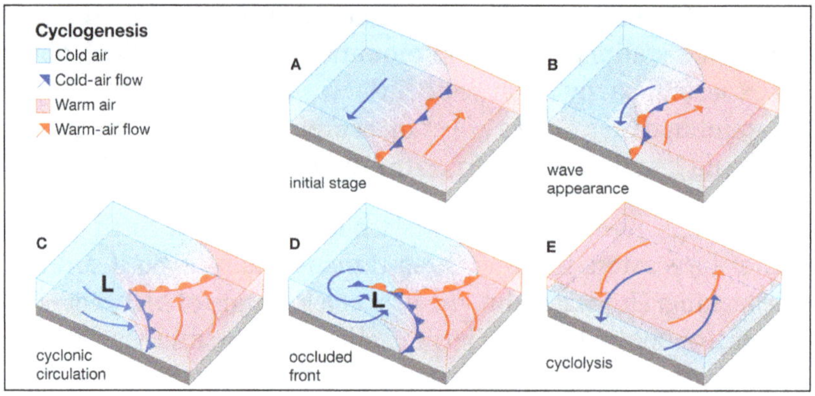

Evolution of a wave (frontal) cyclone

Modern Trends and Developments

Upper-air Observations by Means of Balloon-borne Sounding Equipment

Once again technology provided the means with which to test the new scientific ideas and stimulate yet newer ones. During the late 1920s and '30s, several groups of investigators (those headed by Yrjö Väisälä of Finland and Pavel Aleksandrovich Malchanov of the Soviet Union, for example) began using small radio transmitters with balloon-borne instruments, eliminating the need to recover the instruments and speeding up access to the upper-air data. These radiosondes, as they came to be called, gave rise to the upper-air observation networks that still exist today. Approximately 75 stations in the United States and more than 500 worldwide release, twice daily, balloons that reach heights of 30,000 metres or more. Observations of temperature and relative humidity at various pressures are radioed back to the station from which the balloons are released as they ascend at a predetermined rate. The balloons also are tracked by radar and global positioning system (GPS) satellites to ascertain the behaviour of winds from their drift.

Forecasters are able to produce synoptic weather maps of the upper atmosphere twice each day on the basis of radiosonde observations. While new methods of upper-air measurement have been developed, the primary synoptic clock times for producing upper-air maps are still the radiosonde-observation times—namely, 0000 (midnight) and 1200 (noon) Greenwich Mean Time (GMT). Furthermore, modern computer-based forecasts use 0000 and 1200 GMT as the starting times from which they calculate the changes that are at the heart of modern forecasts. It is, in effect, the synoptic approach carried out in a different way, intimately linked to the radiosonde networks developed during the 1930s and '40s.

Meteorological Measurements from Satellites and Aircraft

A major breakthrough in meteorological measurement came with the launching of the first meteorological satellite, the TIROS (Television and Infrared Observation Satellite), by the United States on April 1, 1960. The impact of global quantitative views of temperature, cloud, and moisture distributions, as well as of surface properties (e.g., ice cover and soil moisture), has already been substantial. Furthermore, new ideas and new methods may very well make the 21st century the "age of the satellite" in weather prediction.

Medium-range forecasts that provide information five to seven days in advance were impossible before satellites began making global observations—particularly over the ocean waters of the Southern Hemisphere—routinely available in real time. Global forecasting models developed at the U.S. National Center for Atmospheric Research (NCAR), the European Centre for Medium Range Weather Forecasts (ECMWF), and the U.S. National Meteorological Center (NMC) became the standard during the 1980s, making medium-range forecasting a reality. Global weather forecasting models are routinely run by national weather services around the world, including those of Japan, the United Kingdom, and Canada.

Meteorological satellites travel in various orbits and carry a wide variety of sensors. They are of two principal types: the low-flying polar orbiter, and the geostationary orbiter.

The first type circle Earth at altitudes of 500–1,000 kilometres and in roughly north–south orbits. They appear overhead at any one locality twice a day and provide very high-resolution data because they fly close to Earth. Such satellites are vitally necessary for much of Europe and other high-latitude locations because they orbit near the poles. These satellites do, however, suffer from one major limitation: they can provide a sampling of atmospheric conditions only twice daily.

The geostationary satellite is made to orbit Earth along its equatorial plane at an altitude of about 36,000 kilometres. At that height the eastward motion of the satellite coincides exactly with Earth's rotation, so that the satellite remains in one position above the Equator. Satellites of this type are able to provide an almost continuous view of a wide area. Because of this capability, geostationary satellites have yielded new information about the rapid changes that occur in thunderstorms, hurricanes, and certain types of fronts, making them invaluable to weather forecasting as well as meteorological research.

One weakness common to virtually all satellite-borne sensors and to some ground-based radars that use UHF/VHF waves is an inability to measure thin layers of the atmosphere. One such layer is the tropopause, the boundary between the relatively dry stratosphere and the more meteorologically active layer below. This is often the region of the jet streams. Important information about these kinds of high-speed air currents is obtained with sensors mounted on high-flying commercial aircraft and is routinely included in global weather analyses.

Short-range Forecasting

Objective Predictions

When people wait under a shelter for a downpour to end, they are making a very-short-range weather forecast. They are assuming, based on past experience, that such hard rain usually does not last very long. In short-term predictions the challenge for the forecaster is to improve on what the layperson can do. For years the type of situation represented in the above example proved particularly vexing for forecasters, but since the mid-1980s they have been developing a method called nowcasting to meet precisely this sort of challenge. In this method, radar and satellite observations of local atmospheric conditions are processed and displayed rapidly by computers to project weather several hours in advance. The U.S. National Oceanic and Atmospheric Administration operates a facility known as PROFS (Program for Regional Observing and Forecasting Services) in Boulder, Colo., specially equipped for nowcasting.

Meteorologists can make somewhat longer-term forecasts (those for 6, 12, 24, or even 48 hours) with considerable skill because they are able to measure and predict atmospheric conditions for large areas by computer. Using models that apply their accumulated expert knowledge quickly, accurately, and in a statistically valid form, meteorologists are now capable of making forecasts objectively. As a consequence, the same results are produced time after time from the same data inputs, with all analysis accomplished mathematically. Unlike the prognostications of the past made with subjective methods, objective forecasts are consistent and can be studied, reevaluated, and improved.

Another technique for objective short-range forecasting is called MOS (for Model Output Statistics). Conceived by Harry R. Glahn and D.A. Lowry of the U.S. National Weather Service, this method involves the use of data relating to past weather phenomena and developments to extrapolate the values of certain weather elements, usually for a specific location and time period. It overcomes the weaknesses of numerical models by developing statistical relations between model forecasts and observed weather. These relations are then used to translate the model forecasts directly to specific weather forecasts. For example, a numerical model might not predict the occurrence of surface winds at all, and whatever winds it did predict might always be too strong. MOS relations can automatically correct for errors in wind speed and produce quite accurate forecasts of wind occurrence at a specific point, such as Heathrow Airport near London. As long as numerical weather prediction models are imperfect, there may be many uses for the MOS technique.

Predictive Skills and Procedures

Short-range weather forecasts generally tend to lose accuracy as forecasters attempt to look farther ahead in time. Predictive skill is greatest for periods of about 12 hours and is still quite substantial for 48-hour predictions. An increasingly important group of short-range forecasts are economically motivated. Their reliability is determined in the marketplace by the economic gains they produce (or the losses they avert).

Weather warnings are a special kind of short-range forecast; the protection of human life is the forecaster's greatest challenge and source of pride. The first national weather forecasting service in the United States (the predecessor of the Weather Bureau) was in fact formed, in 1870, in response to the need for storm warnings on the Great Lakes. Increase Lapham of Milwaukee urged Congress to take action to reduce the loss of hundreds of lives incurred each year by Great Lakes shipping during the 1860s. The effectiveness of the warnings and other forecasts assured the future of the American public weather service.

Weather warnings are issued by government and military organizations throughout the world for all kinds of threatening weather events: tropical storms variously called hurricanes, typhoons, or tropical cyclones, depending on location; great oceanic gales outside the tropics spanning hundreds of kilometres and at times packing winds comparable to those of tropical storms; and, on land, flash floods, high winds, fog, blizzards, ice, and snowstorms.

A particular effort is made to warn of hail, lightning, and wind gusts associated with severe thunderstorms, sometimes called severe local storms (SELS) or simply severe weather. Forecasts and warnings also are made for tornadoes, those intense, rotating windstorms that represent the most violent end of the weather scale. Destruction of property and the risk of injury and death are extremely high in the path of a tornado, especially in the case of the largest systems (sometimes called maxi-tornadoes).

Because tornadoes are so uniquely life-threatening and because they are so common in various regions of the United States, the National Weather Service operates a National Severe Storms Forecasting Center (NSSFC) in Kansas City, Mo., where SELS forecasters survey the atmosphere for the conditions that can spawn tornadoes or severe thunderstorms. This group of SELS forecasters, assembled in 1952, monitors temperature and water vapour in an effort to identify the warm, moist regions where thunderstorms may form and studies maps of pressure and winds to find regions where the storms may organize into mesoscale structures. The group also monitors jet streams and dry air aloft that can combine to distort ordinary thunderstorms into rare rotating ones with tilted chimneys of upward rushing air that, because of the tilt, are unimpeded by heavy falling rain. These high-speed updrafts can quickly transport vast quantities of moisture to the cold upper regions of the storms, thereby promoting the formation of large hailstones. The hail and rain drag down air from aloft to complete a circuit of violent, cooperating updrafts and downdrafts.

By correctly anticipating such conditions, SELS forecasters are able to provide time for the mobilization of special observing networks and personnel. If the storms actually develop, specific warnings are issued based on direct observations. This two-step process consists of the tornado or severe thunderstorm watch, which is the forecast prepared by the SELS forecaster, and the warning, which is usually released by a local observing facility. The watch may be issued when the skies are clear, and it usually covers a number of counties. It alerts the affected area to the threat but does not attempt to pinpoint which communities will be affected.

By contrast, the warning is very specific to a locality and calls for immediate action. Radar of various types can be used to detect the large hailstones, the heavy load of raindrops, the relatively clear region of rapid updraft, and even the rotation in a tornado. These indicators, or an actual sighting, often trigger the tornado warning. In effect, a warning is a specific statement that danger is imminent, whereas a watch is a forecast that warnings may be necessary later in a given region.

Long-range Forecasting

Techniques

Extended-range, or long-range, weather forecasting has had a different history and a different approach from short- or medium-range forecasting. In most cases, it has not applied the synoptic method of going forward in time from a specific initial map. Instead, long-range forecasters have tended to use the climatological approach, often concerning themselves with the broad weather picture over a period of time rather than attempting to forecast day-to-day details.

There is good reason to believe that the limit of day-to-day forecasts based on the "initial map" approach is about two weeks. Most long-range forecasts thus attempt to predict the departures from normal conditions for a given month or season. Such departures are called anomalies. A forecast might state that "spring temperatures in Minneapolis have a 65 percent probability of being above normal." It would likely be based on a forecast anomaly map, which shows temperature anomaly patterns. The maps do not attempt to predict the weather for a particular day, but rather forecast trends (i.e., warmer than normal) for an extended amount of time, such as a season (i.e., spring).

The U.S. Weather Bureau began making experimental long-range forecasts just before the beginning of World War II, and its successor, the National Weather Service, continues to express such predictions in probabilistic terms, making it clear that they are subject to uncertainty. Verification shows that forecasts of temperature anomalies are more reliable than those of precipitation, that monthly forecasts are better than seasonal ones, and that winter months are predicted somewhat more accurately than other seasons.

Prior to the 1980s the technique commonly used in long-range forecasting relied heavily on the analog method, in which groups of weather situations (maps) from previous years were compared to those of the current year to determine similarities with the atmosphere's present patterns (or "habits"). An association was then made between what had happened subsequently in those "similar" years and what was going to happen in the current year. Most of the techniques were quite subjective, and there were often disagreements of interpretation and consequently uneven quality and marginal reliability.

Persistence (warm summers follow warm springs) or anti-persistence (cold springs follow warm winters) also were used, even though, strictly speaking, most forecasters consider persistence forecasts "no-skill" forecasts. Yet, they too have had limited success.

Prospects for new Procedures

In the last quarter of the 20th century the approach of and prospects for long-range weather forecasting changed significantly. Stimulated by the work of Jerome Namias, who headed the U.S. Weather Bureau's Long-Range Forecast Division for 30 years, scientists began to look at ocean-surface temperature anomalies as a potential cause for the temperature anomalies of the atmosphere in succeeding seasons and at distant locations. At the same time, other American meteorologists, most notably John M. Wallace, showed how certain repetitive patterns of atmospheric flow were related to each other in different parts of the world. With satellite-based observations available, investigators began to study the El Niño phenomenon. Atmospheric scientists also revived the work of Gilbert Walker, an early 20th-century British climatologist who had studied the Southern Oscillation, the aforementioned up-and-down fluctuation of atmospheric pressure in the Southern Hemisphere. Walker had investigated related air circulations (later called the Walker Circulation) that resulted from abnormally high pressures in Australia and low pressures in Argentina or vice versa.

All of this led to new knowledge about how the occurrence of abnormally warm or cold ocean waters and of abnormally high or low atmospheric pressures could be interrelated in vast global connections. Knowledge about these links—El Niño/Southern Oscillation (ENSO)—and about the behaviour of parts of these vast systems enables forecasters to make better long-range predictions, at least in part, because the ENSO features change slowly and somewhat regularly. This approach of studying interconnections between the atmosphere and the ocean may represent the beginning of a revolutionary stage in long-range forecasting.

Since the mid-1980s, interest has grown in applying numerical weather prediction models to long-range forecasting. In this case, the concern is not with the details of weather predicted 20 or 30 days in advance but rather with objectively predicted anomalies. The reliability of long-range forecasts, like that of short- and medium-range projections, has improved substantially in recent years. Yet, many significant problems remain unsolved, posing interesting challenges for all those engaged in the field.

WEATHER MAP

A weather map displays various meteorological features across a particular area at a particular point in time and has various symbols which all have specific meanings. Such maps have been in use since the mid-19th century and are used for research and weather forecasting purposes. Maps using isotherms show temperature gradients, which can help locate weather fronts. Isotach maps, analyzing lines of equal wind speed, on a constant pressure surface of 300 or 250 hPa show where the jet stream is located. Use of constant pressure charts at the 700 and 500 hPa level can indicate tropical cyclone motion. Two-dimensional streamlines based on wind speeds at various levels show areas of convergence and divergence in the wind field, which are helpful in determining the location of features within the wind pattern. A popular type of surface weather map is the surface weather analysis, which plots isobars to depict areas of high pressure and low pressure. Cloud codes are translated into symbols and plotted on these maps along with other meteorological data that are included in synoptic reports sent by professionally trained observers.

Plotting of Data

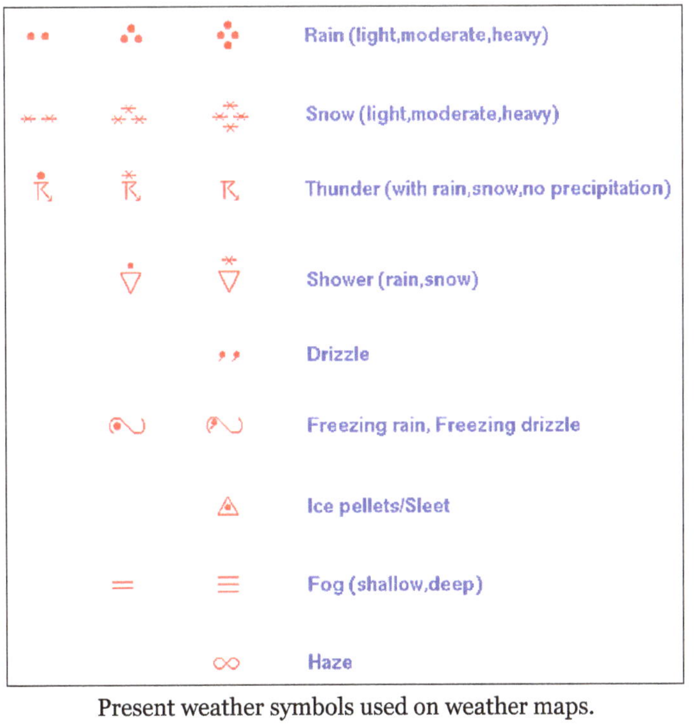

Present weather symbols used on weather maps.

A station model is a symbolic illustration showing the weather occurring at a given reporting station. Meteorologists created the station model to plot a number of weather elements in a small space on weather maps. Maps filled with dense station-model plots can be difficult to read, but they allow meteorologists, pilots, and mariners to see important weather patterns. A computer draws a station model for each observation location. The station model is primarily used on surface-weather maps, but can also be used to show the weather aloft. A completed station-model map allows users to analyze patterns in air pressure, temperature, wind, cloud cover, and precipitation.

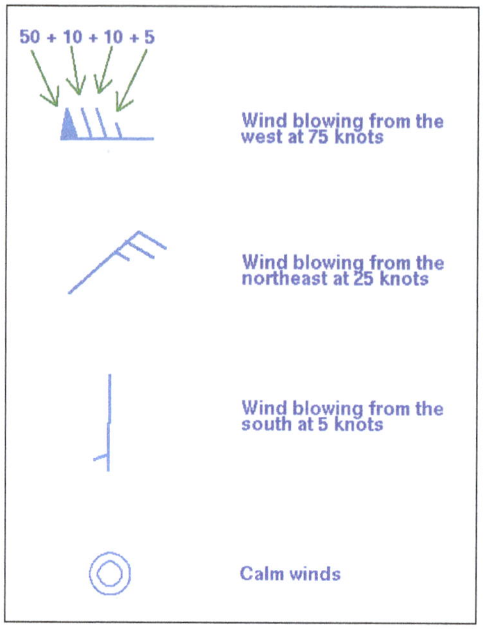

Wind barb interpretation

Station model plots use an internationally accepted coding convention that has changed little since August 1, 1941. Elements in the plot show the key weather elements, including temperature, dewpoint, wind, cloud cover, air pressure, pressure tendency, and precipitation. Winds have a standard notation when plotted on weather maps. More than a century ago, winds were plotted as arrows, with feathers on just one side depicting five knots of wind, while feathers on both sides depicted 10 knots (19 km/h) of wind. The notation changed to that of half of an arrow, with half of a wind barb indicating five knots, a full barb ten knots, and a pennant flag fifty knots.

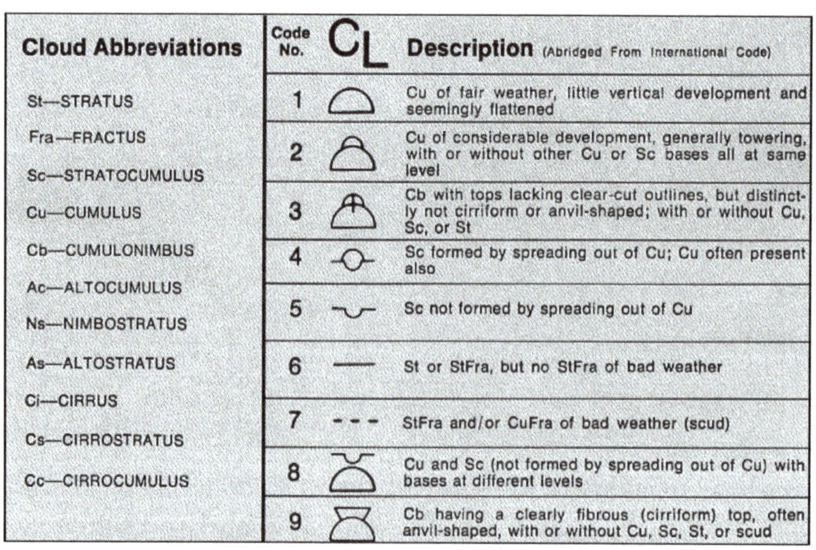

Cloud Abbreviations	Code No.	C_L	Description (Abridged From International Code)
St—STRATUS	1		Cu of fair weather, little vertical development and seemingly flattened
Fra—FRACTUS Sc—STRATOCUMULUS	2		Cu of considerable development, generally towering, with or without other Cu or Sc bases all at same level
Cu—CUMULUS	3		Cb with tops lacking clear-cut outlines, but distinctly not cirriform or anvil-shaped; with or without Cu, Sc, or St
Cb—CUMULONIMBUS	4		Sc formed by spreading out of Cu; Cu often present also
Ac—ALTOCUMULUS Ns—NIMBOSTRATUS	5		Sc not formed by spreading out of Cu
As—ALTOSTRATUS	6		St or StFra, but no StFra of bad weather
Ci—CIRRUS Cs—CIRROSTRATUS	7		StFra and/or CuFra of bad weather (scud)
Cc—CIRROCUMULUS	8		Cu and Sc (not formed by spreading out of Cu) with bases at different levels
	9		Cb having a clearly fibrous (cirriform) top, often anvil-shaped, with or without Cu, Sc, St, or scud

Low étage (Sc,St) and upward-growing vertical (Cu, Cb)

Because of the structure of the SYNOP code, a maximum of three cloud symbols can be plotted for each reporting station that appears on the weather map. All cloud types are coded and transmitted by trained observers then plotted on maps as low, middle, or high-étage using special symbols for each major cloud type. Any cloud type with significant vertical extent that can occupy more than one étage is coded as low (cumulus and cumulonimbus) or middle (nimbostratus) depending on the altitude level or étage where it normally initially forms aside from any vertical growth that takes place. The symbol used on the map for each of these étages at a particular observation time is for the genus, species, variety, mutation, or cloud motion that is considered most important according to criteria set out by the World Meteorological Organization (WMO). If these elements for any étage at the time of observation are deemed to be of equal importance, then the type which is predominant in amount is coded by the observer and plotted on the weather map using the appropriate symbol. Special weather maps in aviation show areas of icing and turbulence.

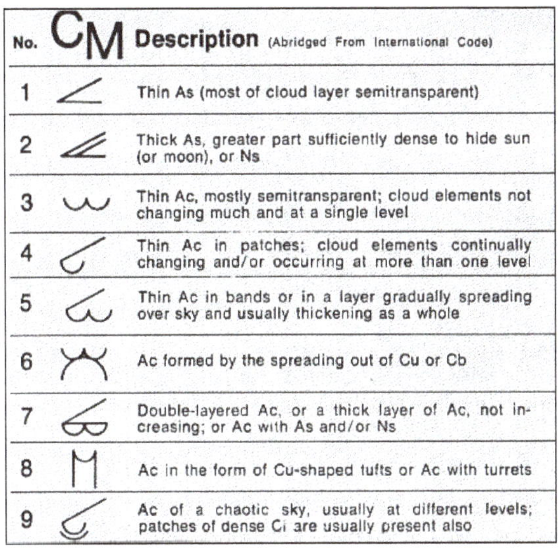

No.	C_M	Description (Abridged From International Code)
1		Thin As (most of cloud layer semitransparent)
2		Thick As, greater part sufficiently dense to hide sun (or moon), or Ns
3		Thin Ac, mostly semitransparent; cloud elements not changing much and at a single level
4		Thin Ac in patches; cloud elements continually changing and/or occurring at more than one level
5		Thin Ac in bands or in a layer gradually spreading over sky and usually thickening as a whole
6		Ac formed by the spreading out of Cu or Cb
7		Double-layered Ac, or a thick layer of Ac, not increasing; or Ac with As and/or Ns
8		Ac in the form of Cu-shaped tufts or Ac with turrets
9		Ac of a chaotic sky, usually at different levels; patches of dense Ci are usually present also

Middle étage (Ac,As) and downward-growing vertical (Ns)

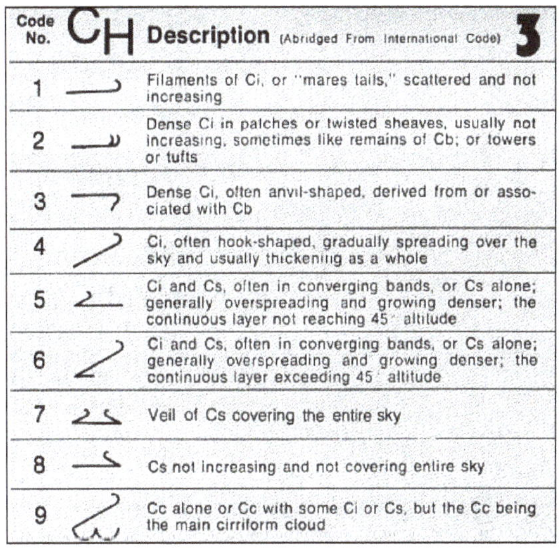

Code No.	C_H	Description (Abridged From International Code) **3**
1		Filaments of Ci, or "mares tails," scattered and not increasing
2		Dense Ci in patches or twisted sheaves, usually not increasing, sometimes like remains of Cb; or towers or tufts
3		Dense Ci, often anvil-shaped, derived from or associated with Cb
4		Ci, often hook-shaped, gradually spreading over the sky and usually thickening as a whole
5		Ci and Cs, often in converging bands, or Cs alone; generally overspreading and growing denser; the continuous layer not reaching 45° altitude
6		Ci and Cs, often in converging bands, or Cs alone; generally overspreading and growing denser; the continuous layer exceeding 45° altitude
7		Veil of Cs covering the entire sky
8		Cs not increasing and not covering entire sky
9		Cc alone or Cc with some Ci or Cs, but the Cc being the main cirriform cloud

High étage (Ci,Cc,Cs)

Types

Aviation Maps

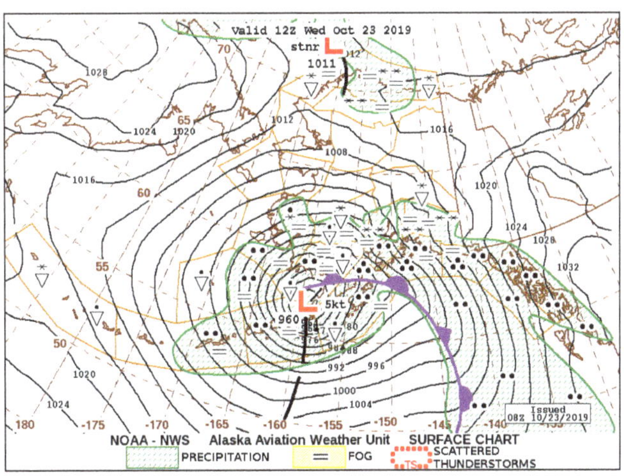

Alaskan aviation weather map

Aviation interests have their own set of weather maps. One type of map shows where VFR (visual flight rules) are in effect and where IFR (instrument flight rules) are in effect. Weather depiction plots show ceiling height (level where at least half the sky is covered with clouds) in hundreds of feet, present weather, and cloud cover. Icing maps depict areas where icing can be a hazard for flying. Aviation-related maps also show areas of turbulence.

Constant Pressure Charts

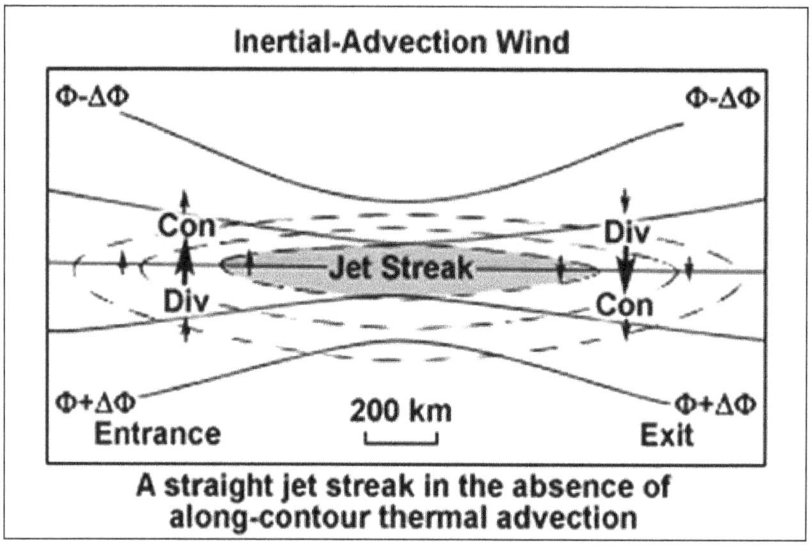

An upper level jet streak. DIV areas are regions of divergence aloft, which usually leads to surface convergence and cyclogenesis

Constant pressure charts normally contain plotted values of temperature, humidity, wind, and the vertical height above sea level of the pressure surface. They have a variety of uses. In the mountainous terrain of the western United States and Mexican Plateau, the 850 hPa pressure surface can be a more realistic depiction of the weather pattern than a standard surface analysis. Using the 850 and 700 hPa pressure surfaces, one can determine when and where warm advection (coincident with upward vertical motion) and cold advection (coincident with downward vertical motion) is occurring within the lower portions of the troposphere. Areas with small dewpoint depressions and are below freezing indicate the presence of icing conditions for aircraft. The 500 hPa pressure surface can be used as a rough guide for the motion of many tropical cyclones. Shallower tropical cyclones, which have experienced vertical wind shear, tend to be steered by winds at the 700 hPa level.

Use of the 300 and 200 hPa constant pressure charts can indicate the strength of systems in the lower troposphere, as stronger systems near the Earth's surface are reflected as stronger features at these levels of the atmosphere. Isotachs are drawn at these levels, which a lines of equal wind speed. They are helpful in finding maxima and minima in the wind pattern. Minima in the wind pattern aloft are favorable for tropical cyclogenesis. Maxima in the wind pattern at various levels of the atmosphere show locations of jet streams. Areas colder than −40 °C (−40 °F) indicate a lack of significant icing, as long as there is no active thunderstorm activity.

A surface weather analysis is a type of weather map that depicts positions for high and low-pressure areas, as well as various types of synoptic scale systems such as frontal zones. Isotherms can be drawn on these maps, which are lines of equal temperature. Isotherms are drawn normally as solid lines at a preferred temperature interval. They show temperature gradients, which can be useful in finding fronts, which are on the warm side of large temperature gradients. By plotting the freezing line, isotherms can be useful in determination of precipitation type. Mesoscale boundaries such as tropical cyclones, outflow boundaries and squall lines also are analyzed on surface weather analyses.

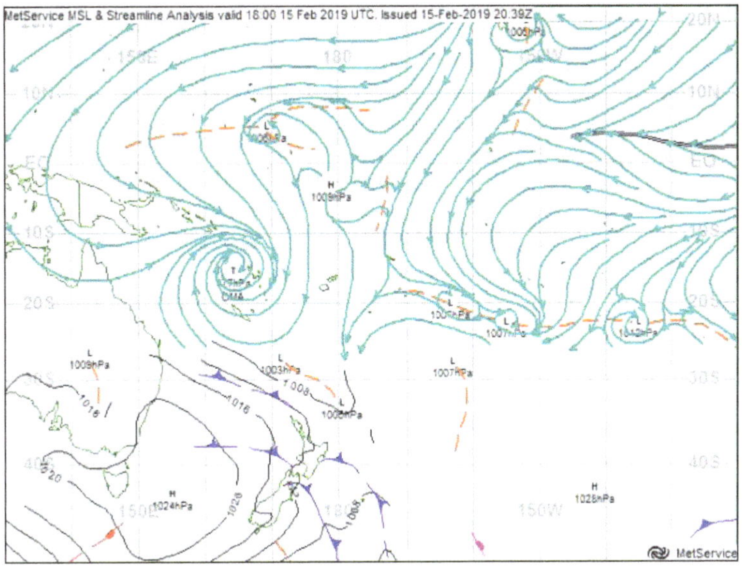

Streamline analysis of the tropical Pacific Ocean

Isobaric analysis is performed on these maps, which involves the construction of lines of equal mean sea level pressure. The innermost closed lines indicate the positions of relative maxima and minima in the pressure field. The minima are called low-pressure areas while the maxima are called high-pressure areas. Highs are often shown as H's whereas lows are shown as L's. Elongated areas of low pressure, or troughs, are sometimes plotted as thick, brown dashed lines down the trough axis. Isobars are commonly used to place surface boundaries from the horse latitudes poleward, while streamline analyses are used in the tropics. A streamline analysis is a series of arrows oriented parallel to wind, showing wind motion within a certain geographic area. C's depict cyclonic flow or likely areas of low pressure, while A's depict anticyclonic flow or likely positions of high-pressure areas. An area of confluent streamlines shows the location of shearlines within the tropics and subtropics.

Surface Weather Analysis

Surface weather analysis is a special type of weather map that provides a view of weather elements over a geographical area at a specified time based on information from ground-based weather stations.

Weather maps are created by plotting or tracing the values of relevant quantities such as sea level pressure, temperature, and cloud cover onto a geographical map to help find synoptic scale features such as weather fronts.

The first weather maps in the 19th century were drawn well after the fact to help devise a theory on storm systems. After the advent of the telegraph, simultaneous surface weather observations became possible for the first time, and beginning in the late 1840s, the Smithsonian Institution became the first organization to draw real-time surface analyses. Use of surface analyses began first in the United States, spreading worldwide during the 1870s. Use of the Norwegian cyclone model for frontal analysis began in the late 1910s across Europe, with its use finally spreading to the United States during World War II.

Surface weather analyses have special symbols that show frontal systems, cloud cover, precipitation, or other important information. For example, an H may represent high pressure, implying good and fair weather. An L, on the other hand, may represent low pressure, which frequently accompanies precipitation. Various symbols are used not just for frontal zones and other surface boundaries on weather maps, but also to depict the present weather at various locations on the weather map. Areas of precipitation help determine the frontal type and location.

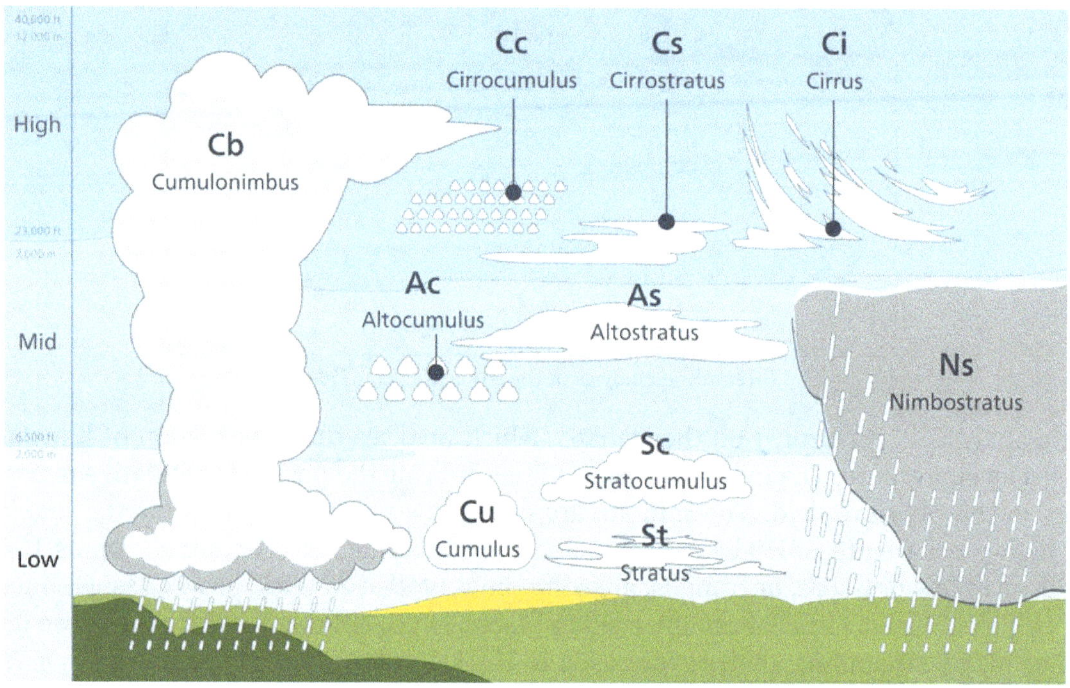

Surface analysis of Great Blizzard of 1888 on March 12, 1888 at 10 pm

The use of weather charts in a modern sense began in the middle portion of the 19th century in order to devise a theory on storm systems. The development of a telegraph network by 1845 made it possible to gather weather information from multiple distant locations quickly enough to preserve its value for real-time applications. The Smithsonian Institution developed its network of observers over much of the central and eastern United States between the 1840s and 1860s once Joseph Henry took the helm. The U.S. Army Signal Corps inherited this network between 1870 and 1874 by an act of Congress, and expanded it to the west coast soon afterwards.

At first, all the data on the map was not taken from these analyses because of a lack of time standardization. The first attempts at time standardization took hold in Great Britain by 1855. The entire United States did not finally come under the influence of time zones until 1905, when Detroit finally established standard time. Other countries followed the lead of the United States in taking simultaneous weather observations, starting in 1873. Other countries then began preparing surface analyses. The use of frontal zones on weather maps did not appear until the introduction of the Norwegian cyclone model in the late 1910s, despite Loomis'earlier attempt at a similar notion in 1841. Since the leading edge of air mass changes bore resemblance to the military fronts of World War I, the term "front" came into use to represent these lines.

Despite the introduction of the Norwegian cyclone model just after World War I, the United States did not formally analyze fronts on surface analyses until late 1942, when the WBAN Analysis Center opened in downtown Washington, D.C. The effort to automate map plotting began in the United States in 1969, with the process complete in the 1970s. Hong Kong completed their process of automated surface plotting by 1987. By 1999, computer systems and software had finally become sophisticated enough to allow for the ability to underlay on the same workstation satellite imagery, radar imagery, and model-derived fields such as atmospheric thickness and frontogenesis in combination with surface observations to make for the best possible surface analysis. In the United States, this development was achieved when Intergraph workstations were replaced by n-AWIPS workstations. By 2001, the various surface analyses done within the National Weather Service were combined into the Unified Surface Analysis, which is issued every six hours and combines the analyses of four different centers. Recent advances in both the fields of meteorology and geographic information systems have made it possible to devise finely tailored products that take us from the traditional weather map into an entirely new realm. Weather information can quickly be matched to relevant geographical detail. For instance, icing conditions can be mapped onto the road network. This will likely continue to lead to changes in the way surface analyses are created and displayed over the next several years. The pressureNET project is an ongoing attempt to gather surface pressure data using smartphones.

Station Model used on Weather Maps

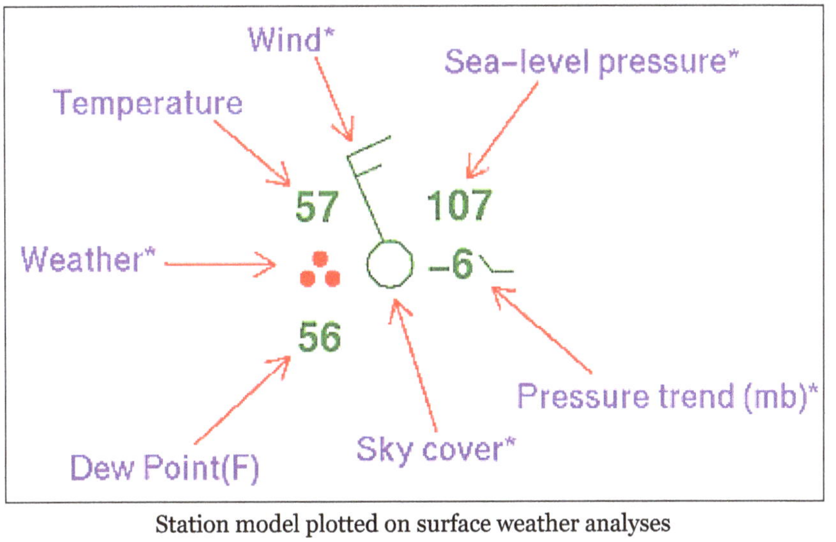

Station model plotted on surface weather analyses

When analyzing a weather map, a station model is plotted at each point of observation. Within the station model, the temperature, dewpoint, wind speed and direction, atmospheric pressure, pressure tendency, and ongoing weather are plotted. The circle in the middle represents cloud cover. If completely filled in, it is overcast. If conditions are completely clear, the circle is empty. If conditions are partly cloudy, the circle is partially filled in. Outside the United States, temperature and dewpoint are plotted in degrees Celsius. Each full flag on the Wind Barb represents 10 knots (19 km/h) of wind, each half flag represents 5 knots (9 km/h). When winds reach 50 knots (93 km/h), a filled in triangle is used for each 50 knots (93 km/h) of wind. In the United States, rainfall plotted in the corner of the station model are in English units, inches. The international standard rainfall measurement unit is the millimeter. Once a map has a field of station models plotted, the analyzing isobars (lines of equal pressure), isallobars (lines of equal pressure change), isotherms (lines of equal temperature), and isotachs (lines of equal wind speed) can be easily accomplished. The abstract present weather symbols used on surface weather analyses for obstructions to visibility, precipitation, and thunderstorms were devised to take up the least room possible on weather maps.

Synoptic Scale Features

A synoptic scale feature is one whose dimensions are large in scale, more than several hundred kilometers in length. Migratory pressure systems and frontal zones exist on this scale.

Pressure Centers

Centers of surface high- and low-pressure areas are found within closed isobars on a surface weather analysis where they are the absolute maxima and minima in the pressure field, and can tell a user in a glance what the general weather is in their vicinity. Weather maps in English-speaking countries will depict their highs as Hs and lows as Ls, while Spanish-speaking countries will depict their highs as As and lows as Bs.

Low Pressure

Low-pressure systems, also known as cyclones, are located in minima in the pressure field. Rotation is inward and counterclockwise in the northern hemisphere as opposed to inward and clockwise in the southern hemisphere due to the Coriolis force. Weather is normally unsettled in the vicinity of a cyclone, with increased cloudiness, increased winds, increased temperatures, and upward motion in the atmosphere, which leads to an increased chance of precipitation. Polar lows can form over relatively mild ocean waters when cold air sweeps in from the ice cap, leading to upward motion and convection, usually in the form of snow. Tropical cyclones and winter storms are intense varieties of low pressure. Over land, thermal lows are indicative of hot weather during the summer.

High Pressure

High-pressure systems, also known as anticyclones, rotate outward and clockwise in the northern hemisphere as opposed to outward and counterclockwise in the southern hemisphere. Under surface highs, sinking motion leads to skies that are clearer, winds that are lighter, and there is a reduced chance of precipitation. There is normally a greater range between high and low temperature due to the drier air mass present. If high pressure persists, air pollution will build up due to pollutants trapped near the surface caused by the subsiding motion associated with the high.

Fronts

Fronts in meteorology are the leading edges of air masses with different density (e.g., air temperature and/or humidity). When a front passes over an area, it is marked by changes in temperature, moisture, wind speed and direction, atmospheric pressure, and often a change in the precipitation pattern. Cold fronts are closely associated with low pressure systems, normally lying at the leading edge of high-pressure systems and, in the case of the polar front, at approximately the equatorward edge of the high-level polar jet. Fronts are guided by winds aloft, but they normally move at lesser speeds. In the northern hemisphere, they usually travel from west to east (though they can move in a north-south direction as well). Movement is due to the pressure gradient force (horizontal differences in atmospheric pressure) and the Coriolis effect, caused by the Earth spinning about its axis. Frontal zones can be contorted by geographic features like mountains and large bodies of water.

Cold Front

A cold front's location is at the leading edge of the temperature drop-off, which in an isotherm analysis shows up as the leading edge of the isotherm gradient, and it normally lies within a sharp surface trough. Cold fronts can move up to twice as fast as warm fronts and produce sharper changes in weather, since cold air is denser than warm air and rapidly replaces the warm air preceding the boundary. Cold fronts are typically accompanied by a narrow band of showers and thunderstorms. On weather maps, the surface position of the cold front is marked with the symbol of a blue line of triangles/spikes (pips) pointing in the direction of travel, and it is placed at the leading edge of the cooler air mass.

Warm Front

Clouds overriding a warm front

Warm fronts are at the trailing edge of the temperature increase, which is located on the equatorward edge of the gradient in isotherms, and lie within broader troughs of low pressure than cold fronts. Warm fronts move more slowly than the cold front that usually follows because cold air is denser, and harder to displace from the Earth's surface. This also forces temperature differences across warm fronts to be broader in scale. Clouds ahead of the warm front are mostly stratiform and rainfall gradually increases as the front approaches. Fog can also occur preceding a warm frontal passage. Clearing and warming is usually rapid after frontal passage. If the warm air mass is unstable, thunderstorms may be embedded among the stratiform clouds ahead of the front, and after frontal passage, thundershowers may continue. On weather maps, the surface location of a warm front is marked with a red line of half circles pointing in the direction of travel.

Occluded Front

An occluded front is formed during the process of cyclogenesis when a cold front overtakes a warm front. The cold and warm fronts curve naturally poleward into the point of occlusion, which is also known as the triple point in meteorology. It lies within a sharp trough, but the air mass behind the boundary can be either warm or cold. In a cold occlusion, the air mass overtaking the warm front is cooler than the cool air ahead of the warm front, and plows under both air masses. In a warm occlusion, the air mass overtaking the warm front is not as cool as the cold air ahead of the warm front, and rides over the colder air mass while lifting the warm air. A wide variety of weather can be found along an occluded front, with thunderstorms possible, but usually their passage is associated with a drying of the air mass. Occluded fronts are indicated on a weather map by a purple line with alternating half-circles and triangles pointing in direction of travel. Occluded fronts usually form around mature low pressure areas.

The trowal is the projection on the Earth's surface of a tongue of warm air aloft, such as may be formed during the occlusion process of a depression.

Stationary Fronts and Shearlines

A stationary front is a non-moving boundary between two different air masses, neither of which is strong enough to replace the other. They tend to remain in the same area for long periods of time, usually moving in waves. There is normally a broad temperature gradient behind the boundary with more widely spaced isotherm packing. A wide variety of weather can be found along a stationary front, but usually clouds and prolonged precipitation are found there. Stationary fronts will either dissipate after several days or devolve into shear lines, but can change into a cold or warm front if conditions aloft change. Stationary fronts are marked on weather maps with alternating red half-circles and blue spikes pointing in opposite directions, indicating no significant movement.

When stationary fronts become smaller in scale, degenerating to a narrow zone where wind direction changes over a short distance, they become known as shear lines. If the shear line becomes active with thunderstorms, it may support formation of a tropical storm or a regeneration of the feature back into a stationary front. A shear line is depicted as a line of red dots and dashes.

Mesoscale Features

Mesoscale features are smaller than synoptic scale systems like fronts, but larger than storm-scale systems like thunderstorms. Horizontal dimensions generally range from over ten kilometres to several hundred kilometres.

Dry Line

The dry line is the boundary between dry and moist air masses east of mountain ranges with similar orientation to the Rockies, depicted at the leading edge of the dew point, or moisture, gradient. Near the surface, warm moist air that is denser than dry air of greater temperature wedges under the drier air like a cold front. When the warm moist air wedged under the drier mass heats up it becomes less dense than the drier air above and it begins to rise and sometimes forms thunderstorms. At higher altitudes, the warm moist air is less dense than the cooler, drier air and the boundary slope reverses. In the vicinity of the reversal aloft, severe weather is possible, especially when a triple point is formed with a cold front.

During daylight hours, drier air from aloft drifts down to the surface, causing an apparent movement of the dryline eastward. At night, the boundary reverts to the west as there is no longer any sunshine to help mix the lower atmosphere. If enough moisture converges upon the dryline, it can be the focus of afternoon and evening thunderstorms. A dry line is depicted on United States surface analyses as a brown line with scallops, or bumps, facing into the moist sector. Dry lines are one of the few surface fronts where the special shapes along the drawn boundary do not necessarily reflect the boundary's direction of motion.

Outflow Boundaries and Squall Lines

A shelf cloud such as this one can be a sign that a squall is imminent

Organized areas of thunderstorm activity not only reinforce pre-existing frontal zones, but they can outrun cold fronts. This outrunning occurs in a pattern where the upper level jet splits into two streams. The resultant mesoscale convective system (MCS) forms at the point of the upper level split in the wind pattern in the area of best low-level inflow. The convection then moves east and equatorward into the warm sector, parallel to low-level thickness lines. When the convection is strong and linear or curved, the MCS is called a squall line, with the feature placed at the leading edge of the significant wind shift and pressure rise. Even weaker and less organized areas of thunderstorms will lead to locally cooler air and higher pressures, and outflow boundaries exist ahead of this type of activity, "Sqln" or "Squall Line", while outflow boundaries are depicted as troughs with a label of "Outflow Boundary" or "Outflow Bndry"

Sea and Land Breeze Fronts

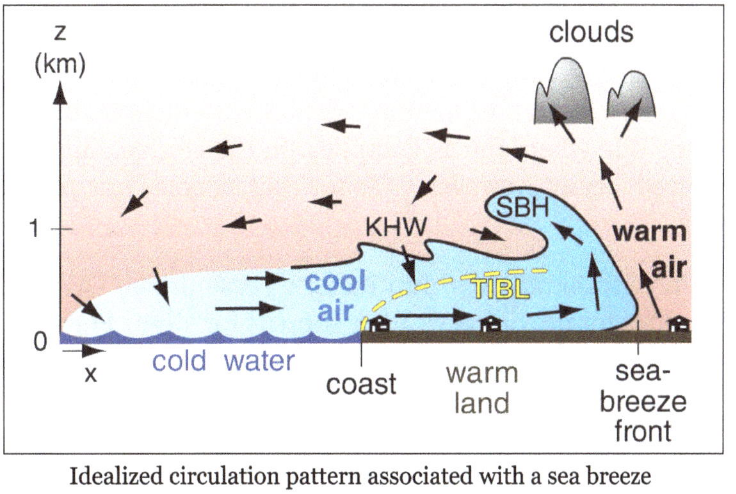

Idealized circulation pattern associated with a sea breeze

Sea breeze fronts occur mainly on sunny days when the landmass warms up above the water temperature. Similar boundaries from downwind on lakes and rivers during the day, as well as offshore landmasses at night. Since the specific heat of water is so high, there is little diurnal change in bodies of water, even on the sunniest days. The water temperature varies less than 1 °C (1 to 2 °F). By contrast, the land, with a lower specific heat, can vary several degrees in a matter of hours.

During the afternoon, air pressure decreases over the land as temperature rises. The relatively cooler air over the sea rushes in to fill the gap. The result is a relatively cool onshore wind. This process usually reverses at night where the water temperature is higher relative to the landmass, leading to an offshore land breeze. However, if water temperatures are colder than the land at night, the sea breeze may continue, only somewhat abated. This is typically the case along the California coast, for example.

If enough moisture exists, thunderstorms can form along sea breeze fronts that then can send out outflow boundaries. This causes chaotic wind/pressure regimes if the steering flow is light. Like all other surface features, sea breeze fronts lie inside troughs of low pressure.

TERMINAL AERODROME FORECAST

In meteorology and aviation, terminal aerodrome forecast (TAF) is a format for reporting weather forecast information, particularly as it relates to aviation. TAFs are issued at least four times a day, every six hours, for major civil airfields: 0000, 0600, 1200 and 1800 UTC, and generally apply to a 24- or 30-hour period, and an area within approximately five statute miles (8.0 km) (or 5 nautical miles (9.3 km) in Canada) from the center of an airport runway complex. TAFs are issued every three hours for military airfields and some civil airfields and cover a period ranging from 3 hours to 30 hours.

TAFs complement and use similar encoding to METAR reports. They are produced by a human forecaster based on the ground. For this reason there are considerably fewer TAF locations than there are airports for which METARs are available. TAFs can be more accurate than Numerical Weather Forecasts, since they take into account local, small-scale, geographic effects.

In the United States the weather forecasters responsible for the TAFs in their respective areas are located within one of the 122 Weather Forecast Offices operated by the United States' National Weather Service. In contrast, a trend type forecast (TTF), which is similar to a TAF, is always produced by a person on-site where the TTF applies. In the United Kingdom most TAFs at military airfields are produced locally, however TAFs for civil airfields are produced at the Met Office headquarters in Exeter.

The United States Air Force employs active duty enlisted personnel as TAF writers. Air Force weather personnel are responsible for providing weather support for all Air Force and Army operations.

Different countries use different change criteria for their weather groups. In the United Kingdom, TAFs for military airfields use colour states as one of the change criteria. Civil airfields in the UK use slightly different criteria.

Code

This TAF example of a 30-hour TAF, released on November 5, 2008 at 1730 UTC:

```
TAF
KXYZ 051730Z 0518/0624 31008KT 3SM -SHRA BKN020
     FM052300 30006KT 5SM -SHRA OVC030
       PROB30 0604/0606 VRB20G35KT 1SM TSRA BKN015CB
     FM060600 25010KT 4SM -SHRA OVC050
       TEMPO 0608/0611 2SM -SHRA OVC030
     RMK NXT FCST BY 00Z=
```

The first line contains identification and validity times.

- TAF indicates that the following is a terminal aerodrome forecast.
- KXYZ indicates the airport to which the forecast applies (ICAO airport code).
- 051730Z indicates that the report was issued on the 5th of the month at 1730 UTC (also known as Zulu, hence the Z).
- 0518/0624 indicates that the report is valid from the 5th at 1800 UTC until the 6th at 2400 UTC.

The remainder of the first line contain the initial forecast conditions. Variations of the codes used for various weather conditions are many.

- 31008KT indicates that the wind will be from 310° true at 8 knots (9 mph; 4 m/s).
- 3SM -SHRA BKN020 indicates that visibility will be 3 statute miles (5 km) in light (-) showers (SH) of rain (RA), with a broken ceiling (between 5⁄8 and 7⁄8 of the sky covered) at 2,000 feet (600 m) above ground level (AGL).

Each line beginning with FM indicates a rapid change in the weather over a period of less than an hour.

- FM052300 indicates the next period lasts from (FM) the 5th at 2300 UTC to the 6th at 0600 UTC (the effective time on the next FM line). The remainder of the line has similar formatting to the other forecast lines: 30006KT wind from 300° at 6 knots (11 km/h; 6.9 mph), 5SM visibility 5 statute miles (8 km), -SHRA light rain showers, OVC030 overcast at 3,000 feet (900 m).
    - PROB30 indicates a 30% probability of the following temporary conditions on the 6th between 0400 UTC and 0600 UTC: VRB20G35KT wind of variable direction at 20 knots (35 km/h; 25 mph) with gusts up to 35 knots (65 km/h; 40 mph), 1SM visibility 1 statute mile (1.6 km), TSRA thunderstorms and rain, BKN015CB broken cumulonimbus clouds at 1,500 feet (450 m).

- FM060600 indicates that the next period lasts from the 6th at 0600 UTC until 2400 UTC (the end of the forecast period, as there is no following FM): 25010KT wind from 250° at 10 knots (20 km/h; 10 mph), 4SM visibility 4 statute miles (6.5 km), -SHRA light rain showers, OVC050 overcast at 5,000 feet (1,500 m).
    - TEMPO means that these conditions are temporarily replaced between 0800 UTC and 1100 UTC with: 2SM visibility 2 statute miles (3 km), -SHRA light rain showers, OVC030 overcast at 3,000 feet (900 m). (The wind speed remains the same as it is not mentioned.)

The final line is for errata, comments, and remarks.

- RMK NXT FCST BY 00Z indicates a remark that the next forecast will be issued by 0000 UTC.

MARINE WEATHER FORECASTING

Marine weather forecasting is the process by which mariners and meteorological organizations attempt to forecast future weather conditions over the Earth's oceans. Mariners have had rules of thumb regarding the navigation around tropical cyclones for many years, dividing a storm into halves and sailing through the normally weaker and more navigable half of their circulation. Marine weather forecasts by various weather organizations can be traced back to the sinking of the Royal Charter in 1859 and the RMS Titanic in 1912.

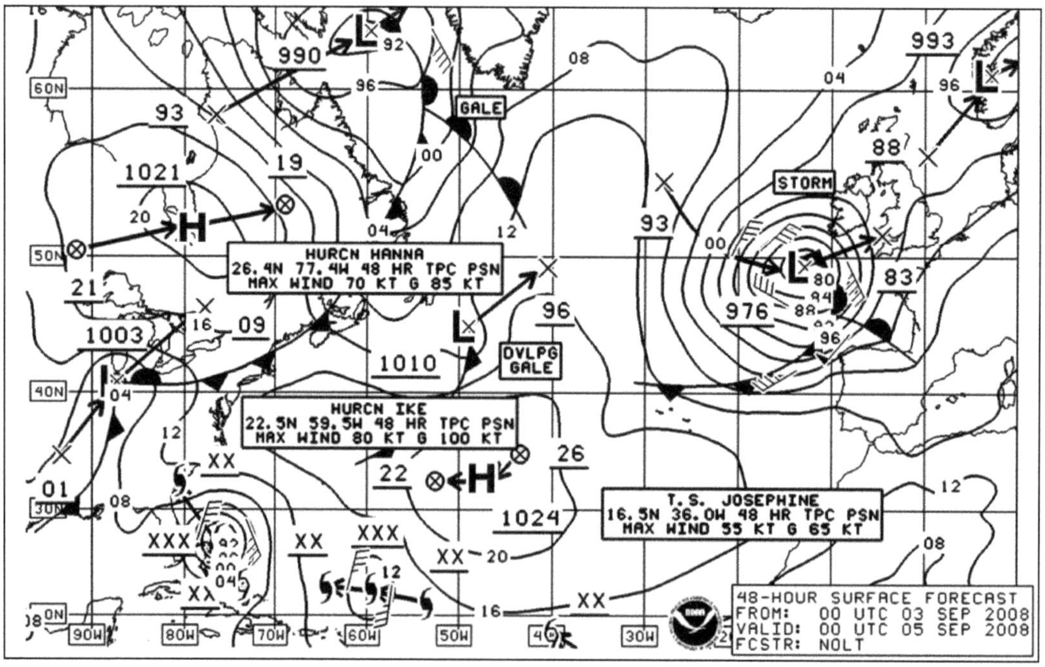

OPC pressure forecast valid at 48 hours

The wind is the driving force of weather at sea, as wind generates local wind waves, long ocean swells, and its flow around the subtropical ridge helps maintain warm water currents such as the

Gulf Stream. The importance of weather over the ocean during World War II led to delayed or secret weather reports, in order to maintain a competitive advantage. Weather ships were established by various nations during World War II for forecasting purposes, and were maintained through 1985 to help with transoceanic plane navigation.

Voluntary observations from ships, weather buoys, weather satellites, and numerical weather prediction have been used to diagnose and help forecast weather over the Earth's ocean areas. Since the 1960s, numerical weather prediction's role over the Earth's seas has taken a greater role in the forecast process. Weather elements such as sea state, surface winds, tide levels, and sea surface temperature are tackled by organizations tasked with forecasting weather over open oceans and seas. Currently, the Japan Meteorological Agency, the United States National Weather Service, and the United Kingdom Met Office create marine weather forecasts for the Northern Hemisphere.

Importance of the Wind

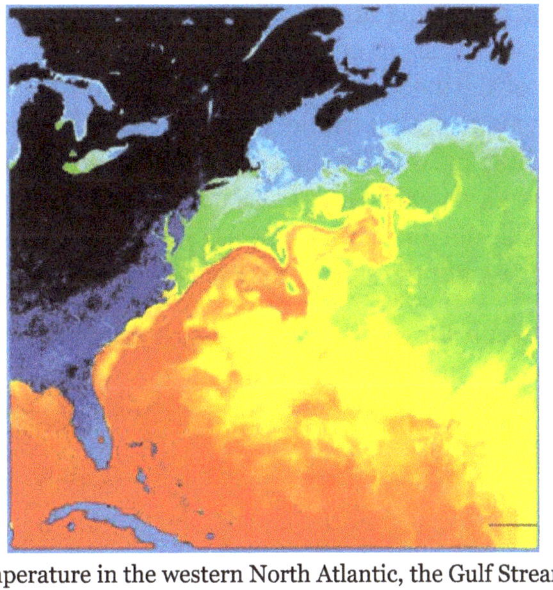

Surface temperature in the western North Atlantic, the Gulf Stream is in red

Development of Warm Ocean Currents

The trade winds blow westward in the tropics, and the westerlies blow eastward at mid-latitudes. This wind pattern applies a stress to the subtropical ocean surface with negative curl across the north Atlantic Ocean. The resulting Sverdrup transport is equatorward. Because of conservation of potential vorticity caused by the poleward-moving winds on the subtropical ridge's western periphery and the increased relative vorticity of northward moving water, transport is balanced by a narrow, accelerating poleward current, which flows along the western boundary of the ocean basin, outweighing the effects of friction with the western boundary current known as the Labrador current. The conservation of potential vorticity also causes bends along the Gulf Stream, which occasionally break off due to a shift in the Gulf Stream's position, forming separate warm and cold eddies. This overall process, known as western intensification, causes currents on the western boundary of an ocean basin, such as the Gulf Stream, to be stronger than those on the eastern boundary.

Swell Dispersion and Wave Groups

North Pacific storm waves as seen from the NOAA M/V *Noble Star*, Winter

Swells are often created by storms long distances away from the beach where they break, and the propagation of the longest swells is only limited by shorelines. For example, swells generated in the Indian Ocean have been recorded in California after more than half a round-the-world trip. This distance allows the waves comprising the swells to be better sorted and free of chop as they travel toward the coast. Waves generated by storm winds have the same speed and will group together and travel with each other, while others moving at even a fraction of a metre per second slower will lag behind, ultimately arriving many hours later due to the distance covered. The time of propagation from the source *t* is proportional to the distance *X* divided by the wave period *T*. In deep water it is where g is the acceleration of gravity. As an example, for a storm located 10,000 kilometres (6,200 mi) away, swells with a period *T*=15 s will arrive 10 days after the storm, followed by 14 s swells another 17 hours later.

This dispersive arrivals of swells, long periods first with a reduction in the peak wave period over time, can be used to tell the distance at which swells were generated. Whereas the sea state in the storm has a frequency spectrum with more or less always the same shape (i.e. a well defined peak with dominant frequencies within plus or minus 7% of the peak), the swell spectra are more and more narrow, sometimes as 2% or less, as waves disperse further and further away. The result is that wave groups (called sets by surfers) can have a large number of waves. From about seven waves per group in the storm, this rises to 20 and more in swells from very distant storms.

Sailing Ship Journeys

Ocean journeys by sailing ship can take many months, and a common hazard is becoming becalmed because of lack of wind, or being blown off course by severe storms or winds that do not allow progress in the desired direction. A severe storm could lead to shipwreck, and the loss of all hands. Sailing ships can only carry a certain quantity of supplies in their hold, so they have to plan long voyages carefully to include appropriate provisions, including fresh water.

Tropical Cyclone Avoidance

Mariners have a way to safely navigate around tropical cyclones. They split tropical cyclones in two, based on their direction of motion, and maneuver to avoid the right segment of the cyclone in the Northern Hemisphere (the left in the Southern Hemisphere). Sailors term the right side the dangerous semicircle since the heaviest rain and strongest winds and seas were located in this half of the storm, as the cyclone's translation speed and its rotational wind are additive. The other half of the tropical cyclone is called the navigable semicircle since weather conditions are lessened (subtractive) in this portion of the storm. The rules of thumb for ship travel when a tropical cyclone is in their vicinity are to avoid them if at all possible and do not cross their forecast path (crossing the T). Those traveling through the dangerous semicircle are advised to keep to the true wind on the starboard bow and make as much headway as possible. Ships moving through the navigable semicircle are advised to keep the true wind on the starboard quarter while making as much headway as possible.

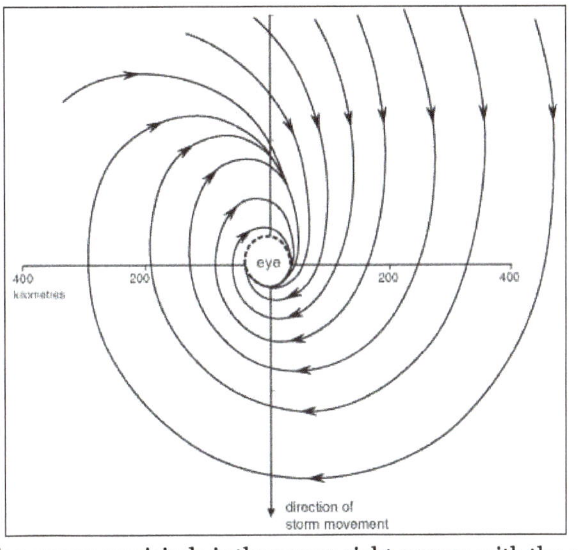

The dangerous semicircle is the upper-right corner, with the arrow marking the direction of motion of a Northern Hemisphere storm

The 1-2-3 rule (mariners' 1-2-3 rule or danger area) is a guideline commonly taught to mariners for severe storm (specifically hurricane and tropical storm) tracking and prediction. It refers to the rounded long-term National Hurricane Center forecast errors of 100-200-300 nautical miles at 24-48-72 hours, respectively. However, these errors have decreased to near 50-100-150 as NHC forecasters become more accurate with tropical cyclone track forecasting. The "danger area" to be avoided is constructed by expanding the forecast path by a radius equal to the respective hundreds of miles plus the forecast wind radii (size of the storm at those hours).

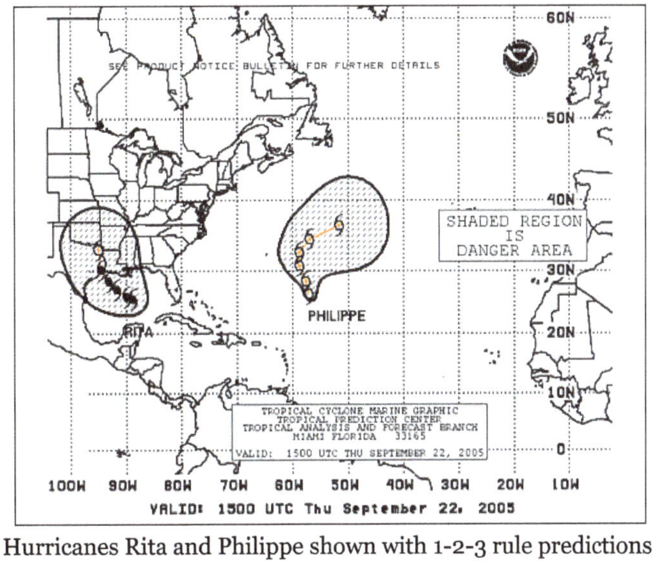

Hurricanes Rita and Philippe shown with 1-2-3 rule predictions

Ocean Surface Modeling

The transfer of energy between the wind blowing over the surface of an ocean and the ocean's upper layer is an important element in wave dynamics. The spectral wave transport equation is used to describe the change in wave spectrum over changing topography. It simulates wave generation, wave movement (propagation within a fluid), wave shoaling, refraction, energy transfer between waves, and wave dissipation. Since surface winds are the primary forcing mechanism in the spectral wave transport equation, ocean wave models use information produced by numerical weather prediction models as inputs to determine how much energy is transferred from the atmosphere into the layer at the surface of the ocean. Along with dissipation of energy through whitecaps and resonance between waves, surface winds from numerical weather models allow for more accurate predictions of the state of the sea surface.

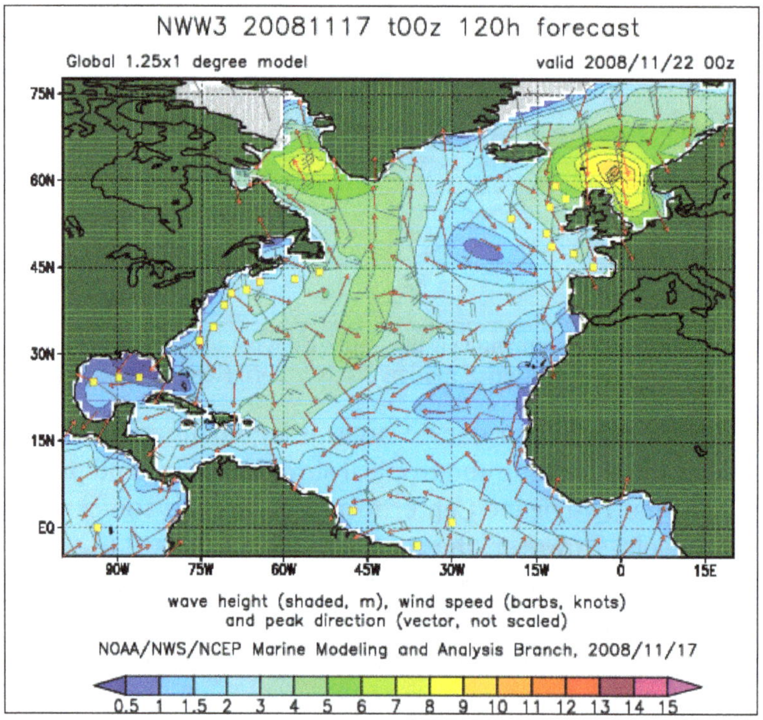

120-hour wind and wave forecast for the North Atlantic

The first ocean wave models were developed in the 1960s and 1970s. These models had the tendency to overestimate the role of wind in wave development and underplayed wave interactions. A lack of knowledge concerning how waves interacted among each other, assumptions regarding a maximum wave height, and deficiencies in computer power limited the performance of the models. After experiments were performed in 1968, 1969, and 1973, wind input from the Earth's atmosphere was weighted more accurately in the predictions. A second generation of models was developed in the 1980s, but they could not realistically model swell nor depict wind-driven waves (also known as wind waves) caused by rapidly changing wind fields, such as those within tropical cyclones. This caused the development of a third generation of wave models from 1988 onward.

Within this third generation of models, the spectral wave transport equation is used to describe the change in wave spectrum over changing topography. It simulates wave generation, wave movement (propagation within a fluid), wave shoaling, refraction, energy transfer between waves, and wave dissipation. Since surface winds are the primary forcing mechanism in the spectral wave transport equation, ocean wave models use information produced by numerical weather prediction models as inputs to determine how much energy is transferred from the atmosphere into the layer at the surface of the ocean. Along with dissipation of energy through whitecaps and resonance between waves, surface winds from numerical weather models allow for more accurate predictions of the state of the sea surface.

Observing Platforms

Weather Ships

The idea of a stationary weather ship was proposed as early as 1921 by Météo-France to help support shipping and the coming of transatlantic aviation. Established during World War II, a weather ship, or ocean weather vessel, was a ship stationed in the ocean as a platform for surface and upper air meteorological observations for use in weather forecasting. They were used during World War II but had no means of defense, which led to the loss of several ships and many lives. They were primarily located in the north Atlantic and north Pacific oceans, reporting via radio. In addition to their weather reporting function, these vessels aided in search and rescue operations, supported transatlantic flights, acted as research platforms for oceanographers, monitored marine pollution, and aided weather forecasting both by weather forecasters and within computerized atmospheric models. Research vessels remain heavily used in oceanography, including physical oceanography and the integration of meteorological and climatological data in Earth system science.

The weather ship MS *Polarfront* at sea

The establishment of weather ships proved to be so useful during World War II that the International Civil Aviation Organization (ICAO) had established a global network of 13 weather ships by 1948, with seven operated by the United States, one operated jointly by the United States and Canada, two supplied by the United Kingdom, one maintained by France, one a joint venture by the Netherlands and Belgium, and one shared by the United Kingdom, Norway, and Sweden. This number was eventually negotiated down to nine. The agreement of the use of weather ships by the international community ended in 1985.

Weather buoy operated by the NOAA National Data Buoy Center

Weather Buoys

Weather buoys are instruments which collect weather and ocean data within the world's oceans, as well as aid during emergency response to chemical spills, legal proceedings, and engineering design. Moored buoys have been in use since 1951, while drifting buoys have been used since 1972. Moored buoys are connected with the ocean bottom using either chains, nylon, or buoyant polypropylene. With the decline of the weather ship, they have taken a more primary role in measuring conditions over the open seas since the 1970s. During the 1980s and 1990s, a network of buoys in the central and eastern tropical Pacific Ocean helped study the El Niño-Southern Oscillation. Moored weather buoys range from 1.5 metres (4.9 ft) to 12 metres (39 ft) in diameter, while drifting buoys are smaller, with diameters of 30 centimetres (12 in) to 40 centimetres (16 in). Drifting buoys are the dominant form of weather buoy in sheer number, with 1250 located worldwide. Wind data from buoys has smaller error than that from ships. There are differences in the values of sea surface temperature measurements between the two platforms as well, relating to the depth of the measurement and whether or not the water is heated by the ship which measures the quantity.

Weather Satellites

In use since 1960, the weather satellite is a type of satellite that is primarily used to monitor the weather and climate of the Earth. Satellites can be polar orbiting, covering the entire Earth asynchronously, or geostationary, hovering over the same spot on the equator. Meteorological satellites see more than clouds and cloud systems. Beginning with the Nimbus 3 satellite in 1969, temperature information through the atmospheric column began to be retrieved by satellites from the eastern Atlantic and most of the Pacific Ocean, which led to significant forecast improvements. City lights, fires, effects of pollution, auroras, sand and dust storms, snow cover, ice mapping, boundaries of ocean currents, energy flows, etc., and other types of environmental information are collected using weather satellites. Other environmental satellites can detect changes in the Earth's vegetation, sea state, ocean color, and ice fields. El Niño and its effects on weather are monitored daily from satellite images. Collectively, weather satellites flown by the U.S., Europe, India, China, Russia, and Japan provide nearly continuous observations for a global weather watch.

The first image obtained from the GOES 1 satellite

Utility

Commercial and recreational use of waterways can be limited significantly by wind direction and speed, wave periodicity and heights, tides, and precipitation. These factors can each influence the safety of marine transit. Consequently, a variety of codes have been established to efficiently transmit detailed marine weather forecasts to vessel pilots via radio, for example the MAFOR (marine forecast). Typical weather forecasts can be received at sea through the use of RTTY, Navtex and Radiofax.

Marine weather warnings and forecasts in print and prognostic chart formats are produced for up five days into the future. Forecasts in printed form include the High Seas Forecast, Offshore Marine Forecasts, and Coastal Waters Forecasts. To help shorten the length of the forecast products, single words and phrases are used to describe areas out at sea. Experimental gridded significant wave height forecasts began being produced by the Ocean Prediction Center in 2006, a first step toward digital marine service for high seas and offshore areas. Additional gridded products such as surface pressure and winds are under development. Recently, National Weather Service operational extratropical storm surge model output to provide experimental extratropical storm surge guidance for coastal weather forecast offices to assist them in coastal flood warning and forecast operations.

NUMERICAL WEATHER PREDICTION

Weather models use systems of differential equations based on the laws of physics, which are in detail fluid motion, thermodynamics, radiative transfer, and chemistry, and use a coordinate system which divides the planet into a 3D grid. Winds, heat transfer, solar radiation, relative humidity, phase changes of water and surface hydrology are calculated within each grid cell, and the interactions with neighboring cells are used to calculate atmospheric properties in the future.

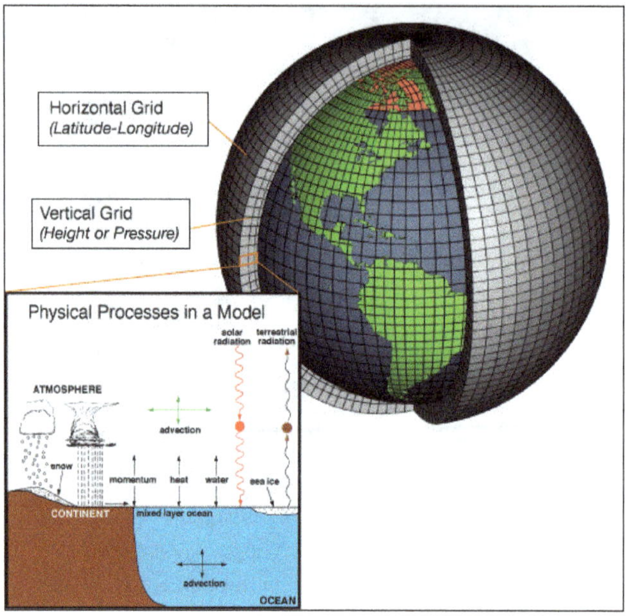

Numerical weather prediction (NWP) uses mathematical models of the atmosphere and oceans to predict the weather based on current weather conditions. Though first attempted in the 1920s, it was not until the advent of computer simulation in the 1950s that numerical weather predictions produced realistic results. A number of global and regional forecast models are run in different countries worldwide, using current weather observations relayed from radiosondes, weather satellites and other observing systems as inputs.

Mathematical models based on the same physical principles can be used to generate either short-term weather forecasts or longer-term climate predictions; the latter are widely applied for understanding and projecting climate change. The improvements made to regional models have allowed for significant improvements in tropical cyclone track and air quality forecasts; however, atmospheric models perform poorly at handling processes that occur in a relatively constricted area, such as wildfires.

Manipulating the vast datasets and performing the complex calculations necessary to modern numerical weather prediction requires some of the most powerful supercomputers in the world. Even with the increasing power of supercomputers, the forecast skill of numerical weather models extends to only about six days. Factors affecting the accuracy of numerical predictions include the density and quality of observations used as input to the forecasts, along with deficiencies in the numerical models themselves. Post-processing techniques such as model output statistics (MOS) have been developed to improve the handling of errors in numerical predictions.

A more fundamental problem lies in the chaotic nature of the partial differential equations that govern the atmosphere. It is impossible to solve these equations exactly, and small errors grow with time (doubling about every five days). Present understanding is that this chaotic behavior limits accurate forecasts to about 14 days even with perfectly accurate input data and a flawless model. In addition, the partial differential equations used in the model need to be supplemented with parameterizations for solar radiation, moist processes (clouds and precipitation), heat exchange, soil, vegetation, surface water, and the effects of terrain. In an effort to quantify the large amount of inherent uncertainty remaining in numerical predictions, ensemble forecasts have been used since the 1990s to help gauge the confidence in the forecast, and to obtain useful results farther into the future than otherwise possible. This approach analyzes multiple forecasts created with an individual forecast model or multiple models.

Initialization

Weather reconnaissance aircraft, such as this WP-3D Orion, provide data that is then used in numerical weather forecasts

The atmosphere is a fluid. As such, the idea of numerical weather prediction is to sample the state of the fluid at a given time and use the equations of fluid dynamics and thermodynamics to estimate the state of the fluid at some time in the future. The process of entering observation data into the model to generate initial conditions is called *initialization*. On land, terrain maps available at resolutions down to 1 kilometer (0.6 mi) globally are used to help model atmospheric circulations within regions of rugged topography, in order to better depict features such as downslope winds, mountain waves and related cloudiness that affects incoming solar radiation. The main inputs from country-based weather services are observations from devices (called radiosondes) in weather balloons that measure various atmospheric parameters and transmits them to a fixed receiver, as well as from weather satellites. The World Meteorological Organization acts to standardize the instrumentation, observing practices and timing of these observations worldwide. Stations either report hourly in METAR reports, or every six hours in SYNOP reports. These observations are irregularly spaced, so they are processed by data assimilation and objective analysis methods, which perform quality control and obtain values at locations usable by the model's mathematical algorithms. The data are then used in the model as the starting point for a forecast.

A variety of methods are used to gather observational data for use in numerical models. Sites launch radiosondes in weather balloons which rise through the troposphere and well into the stratosphere. Information from weather satellites is used where traditional data sources are not available. Commerce provides pilot reports along aircraft routes and ship reports along shipping routes. Research projects use reconnaissance aircraft to fly in and around weather systems of interest, such as tropical cyclones. Reconnaissance aircraft are also flown over the open oceans during the cold season into systems which cause significant uncertainty in forecast guidance, or are expected to be of high impact from three to seven days into the future over the downstream continent. Sea ice began to be initialized in forecast models in 1971. Efforts to involve sea surface temperature in model initialization began in 1972 due to its role in modulating weather in higher latitudes of the Pacific.

Computation

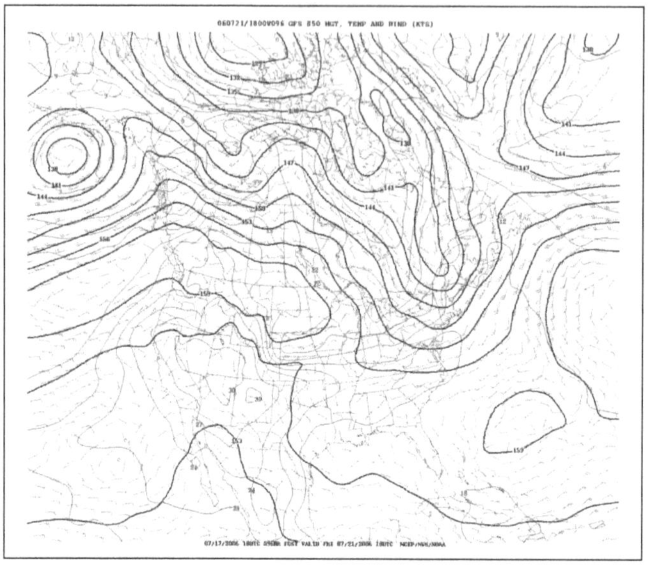

A prognostic chart of the 96-hour forecast of 850 mbar geopotential height and temperature from the Global Forecast System

An atmospheric model is a computer program that produces meteorological information for future times at given locations and altitudes. Within any modern model is a set of equations, known as the primitive equations, used to predict the future state of the atmosphere. These equations—along with the ideal gas law—are used to evolve the density, pressure, and potential temperature scalar fields and the air velocity (wind) vector field of the atmosphere through time. Additional transport equations for pollutants and other aerosols are included in some primitive-equation high-resolution models as well. The equations used are nonlinear partial differential equations which are impossible to solve exactly through analytical methods, with the exception of a few idealized cases. Therefore, numerical methods obtain approximate solutions. Different models use different solution methods: some global models and almost all regional models use finite difference methods for all three spatial dimensions, while other global models and a few regional models use spectral methods for the horizontal dimensions and finite-difference methods in the vertical.

These equations are initialized from the analysis data and rates of change are determined. These rates of change predict the state of the atmosphere a short time into the future; the time increment for this prediction is called a *time step*. This future atmospheric state is then used as the starting point for another application of the predictive equations to find new rates of change, and these new rates of change predict the atmosphere at a yet further time step into the future. This time stepping is repeated until the solution reaches the desired forecast time. The length of the time step chosen within the model is related to the distance between the points on the computational grid, and is chosen to maintain numerical stability. Time steps for global models are on the order of tens of minutes, while time steps for regional models are between one and four minutes. The global models are run at varying times into the future. The UKMET Unified Model is run six days into the future, while the European Centre for Medium-Range Weather Forecasts' Integrated Forecast System and Environment Canada's Global Environmental Multiscale Model both run out to ten days into the future, and the Global Forecast System model run by the Environmental Modeling Center is run sixteen days into the future. The visual output produced by a model solution is known as a prognostic chart, or *prog*.

Parameterization

Field of cumulus clouds, which are parameterized since they are too small to be explicitly included within numerical weather prediction

Some meteorological processes are too small-scale or too complex to be explicitly included in numerical weather prediction models. *Parameterization* is a procedure for representing these processes by relating them to variables on the scales that the model resolves. For example, the gridboxes in weather and climate models have sides that are between 5 kilometers (3 mi) and 300 kilometers (200 mi) in length. A typical cumulus cloud has a scale of less than 1 kilometer (0.6 mi), and would require a grid even finer than this to be represented physically by the equations of fluid motion. Therefore, the processes that such clouds represent are parameterized, by processes of various sophistication. In the earliest models, if a column of air within a model gridbox was conditionally unstable (essentially, the bottom was warmer and moister than the top) and the water vapor content at any point within the column became saturated then it would be overturned (the warm, moist air would begin rising), and the air in that vertical column mixed. More sophisticated schemes recognize that only some portions of the box might convect and that entrainment and other processes occur. Weather models that have gridboxes with sizes between 5 and 25 kilometers (3 and 16 mi) can explicitly represent convective clouds, although they need to parameterize cloud microphysics which occur at a smaller scale. The formation of large-scale (stratus-type) clouds is more physically based; they form when the relative humidity reaches some prescribed value. Sub-grid scale processes need to be taken into account. Rather than assuming that clouds form at 100% relative humidity, the cloud fraction can be related to a critical value of relative humidity less than 100%, reflecting the sub grid scale variation that occurs in the real world.

The amount of solar radiation reaching the ground, as well as the formation of cloud droplets occur on the molecular scale, and so they must be parameterized before they can be included in the model. Atmospheric drag produced by mountains must also be parameterized, as the limitations in the resolution of elevation contours produce significant underestimates of the drag. This method of parameterization is also done for the surface flux of energy between the ocean and the atmosphere, in order to determine realistic sea surface temperatures and type of sea ice found near the ocean's surface. Sun angle as well as the impact of multiple cloud layers is taken into account. Soil type, vegetation type, and soil moisture all determine how much radiation goes into warming and how much moisture is drawn up into the adjacent atmosphere, and thus it is important to parameterize their contribution to these processes. Within air quality models, parameterizations take into account atmospheric emissions from multiple relatively tiny sources (e.g. roads, fields, factories) within specific grid boxes.

Domains

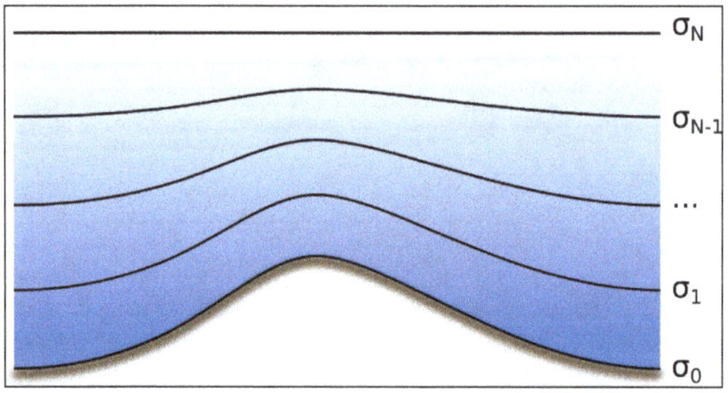

A cross-section of the atmosphere over terrain with a sigma-coordinate representation shown. Mesoscale models divide the atmosphere vertically using representations similar to the one shown here

The horizontal domain of a model is either *global*, covering the entire Earth, or *regional*, covering only part of the Earth. Regional models (also known as *limited-area* models, or LAMs) allow for the use of finer grid spacing than global models because the available computational resources are focused on a specific area instead of being spread over the globe. This allows regional models to resolve explicitly smaller-scale meteorological phenomena that cannot be represented on the coarser grid of a global model. Regional models use a global model to specify conditions at the edge of their domain (boundary conditions) in order to allow systems from outside the regional model domain to move into its area. Uncertainty and errors within regional models are introduced by the global model used for the boundary conditions of the edge of the regional model, as well as errors attributable to the regional model itself.

Coordinate Systems

Horizontal Coordinates

Horizontal position may be expressed directly in geographic coordinates (latitude and longitude) for global models or in a map projection planar coordinates for regional models. The German weather service is using for its global ICON model (icosahedral non-hydrostatic global circulation model) a grid based on a regular icosahedron. Basic cells in this grid are triangles instead of the four corner cells in a traditional latitude-longitude grid. The advantage is that, different from a latitude-longitude cells are everywhere on the globe the same size. Disadvantage is that equations in this non rectangular grid are more complicated.

Vertical Coordinates

The vertical coordinate is handled in various ways. Lewis Fry Richardson's 1922 model used geometric height (z) as the vertical coordinate. Later models substituted the geometric z coordinate with a pressure coordinate system, in which the geopotential heights of constant-pressure surfaces become dependent variables, greatly simplifying the primitive equations. This correlation between coordinate systems can be made since pressure decreases with height through the Earth's atmosphere. The first model used for operational forecasts, the single-layer barotropic model, used a single pressure coordinate at the 500-millibar (about 5,500 m (18,000 ft)) level, and thus was essentially two-dimensional. High-resolution models—also called *mesoscale models*—such as the Weather Research and Forecasting model tend to use normalized pressure coordinates referred to as sigma coordinates. This coordinate system receives its name from the independent variable σ used to scale atmospheric pressures with respect to the pressure at the surface, and in some cases also with the pressure at the top of the domain.

Model Output Statistics

Because forecast models based upon the equations for atmospheric dynamics do not perfectly determine weather conditions, statistical methods have been developed to attempt to correct the forecasts. Statistical models were created based upon the three-dimensional fields produced by numerical weather models, surface observations and the climatological conditions for specific locations. These statistical models are collectively referred to as model output statistics (MOS), and were developed by the National Weather Service for their suite of weather forecasting models in the late 1960s.

Model output statistics differ from the *perfect prog* technique, which assumes that the output of numerical weather prediction guidance is perfect. MOS can correct for local effects that cannot be resolved by the model due to insufficient grid resolution, as well as model biases. Because MOS is run after its respective global or regional model, its production is known as post-processing. Forecast parameters within MOS include maximum and minimum temperatures, percentage chance of rain within a several hour period, precipitation amount expected, chance that the precipitation will be frozen in nature, chance for thunderstorms, cloudiness, and surface winds.

Applications

Air Quality Modeling

Air quality forecasting attempts to predict when the concentrations of pollutants will attain levels that are hazardous to public health. The concentration of pollutants in the atmosphere is determined by their *transport*, or mean velocity of movement through the atmosphere, their diffusion, chemical transformation, and ground deposition. In addition to pollutant source and terrain information, these models require data about the state of the fluid flow in the atmosphere to determine its transport and diffusion. Meteorological conditions such as thermal inversions can prevent surface air from rising, trapping pollutants near the surface, which makes accurate forecasts of such events crucial for air quality modeling. Urban air quality models require a very fine computational mesh, requiring the use of high-resolution mesoscale weather models; in spite of this, the quality of numerical weather guidance is the main uncertainty in air quality forecasts.

Climate Modeling

A General Circulation Model (GCM) is a mathematical model that can be used in computer simulations of the global circulation of a planetary atmosphere or ocean. An atmospheric general circulation model (AGCM) is essentially the same as a global numerical weather prediction model, and some (such as the one used in the UK Unified Model) can be configured for both short-term weather forecasts and longer-term climate predictions. Along with sea ice and land-surface components, AGCMs and oceanic GCMs (OGCM) are key components of global climate models, and are widely applied for understanding the climate and projecting climate change. For aspects of climate change, a range of man-made chemical emission scenarios can be fed into the climate models to see how an enhanced greenhouse effect would modify the Earth's climate. Versions designed for climate applications with time scales of decades to centuries were originally created in 1969 by Syukuro Manabe and Kirk Bryan at the Geophysical Fluid Dynamics Laboratory in Princeton, New Jersey. When run for multiple decades, computational limitations mean that the models must use a coarse grid that leaves smaller-scale interactions unresolved.

Ocean Surface Modeling

The transfer of energy between the wind blowing over the surface of an ocean and the ocean's upper layer is an important element in wave dynamics. The spectral wave transport equation is used to describe the change in wave spectrum over changing topography. It simulates wave generation, wave movement (propagation within a fluid), wave shoaling, refraction, energy transfer between waves, and wave dissipation. Since surface winds are the primary forcing mechanism in the spectral wave transport equation, ocean wave models use information produced by numerical weather prediction models as inputs to determine how much energy is transferred from the atmosphere into the layer at the surface of the ocean. Along with dissipation of energy through whitecaps and resonance between waves, surface winds from numerical weather models allow for more accurate predictions of the state of the sea surface.

Tropical Cyclone Forecasting

Tropical cyclone forecasting also relies on data provided by numerical weather models. Three main classes of tropical cyclone guidance models exist: Statistical models are based on an analysis of storm behavior using climatology, and correlate a storm's position and date to produce a forecast that is not based on the physics of the atmosphere at the time. Dynamical models are numerical models that solve the governing equations of fluid flow in the atmosphere; they are based on the same principles as other limited-area numerical weather prediction models but may include special computational techniques such as refined spatial domains that move along with the cyclone. Models that use elements of both approaches are called statistical-dynamical models.

In 1978, the first hurricane-tracking model based on atmospheric dynamics—the movable fine-mesh (MFM) model—began operating. Within the field of tropical cyclone track forecasting, despite the ever-improving dynamical model guidance which occurred with increased computational power, it was not until the 1980s when numerical weather prediction showed skill, and until the 1990s when it consistently outperformed statistical or simple dynamical models. Predictions of the intensity of a tropical cyclone based on numerical weather prediction continue to be a challenge, since statistical methods continue to show higher skill over dynamical guidance.

Wildfire Modeling

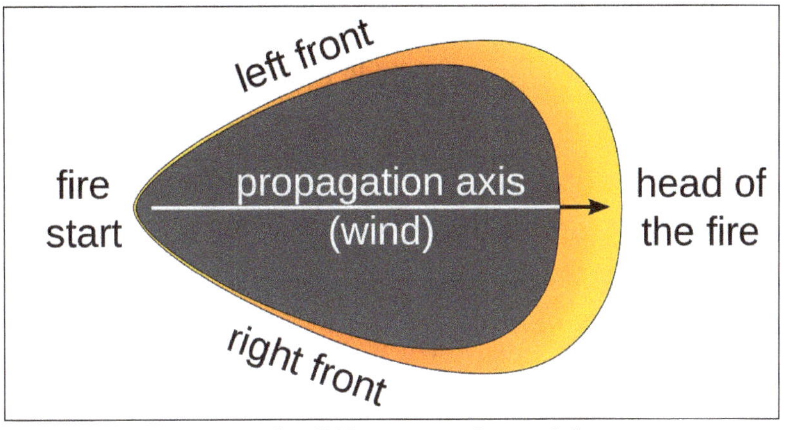

A simple wildfire propagation model

On a molecular scale, there are two main competing reaction processes involved in the degradation of cellulose, or wood fuels, in wildfires. When there is a low amount of moisture in a cellulose fiber, volatilization of the fuel occurs; this process will generate intermediate gaseous products that will ultimately be the source of combustion. When moisture is present—or when enough heat is being carried away from the fiber, charring occurs. The chemical kinetics of both reactions indicate that there is a point at which the level of moisture is low enough—and/or heating rates high enough—for combustion processes become self-sufficient. Consequently, changes in wind speed, direction, moisture, temperature, or lapse rate at different levels of the atmosphere can have a significant impact on the behavior and growth of a wildfire. Since the wildfire acts as a heat source to the atmospheric flow, the wildfire can modify local advection patterns, introducing a feedback loop between the fire and the atmosphere.

A simplified two-dimensional model for the spread of wildfires that used convection to represent the effects of wind and terrain, as well as radiative heat transfer as the dominant method of heat transport led to reaction-diffusion systems of partial differential equations. More complex models join numerical weather models or computational fluid dynamics models with a wildfire component which allow the feedback effects between the fire and the atmosphere to be estimated. The additional complexity in the latter class of models translates to a corresponding increase in their computer power requirements. In fact, a full three-dimensional treatment of combustion via direct numerical simulation at scales relevant for atmospheric modeling is not currently practical because of the excessive computational cost such a simulation would require. Numerical weather models have limited forecast skill at spatial resolutions under 1 kilometer (0.6 mi), forcing complex wildfire models to parameterize the fire in order to calculate how the winds will be modified locally by the wildfire, and to use those modified winds to determine the rate at which the fire will spread locally. Although models such as Los Alamos' FIRETEC solve for the concentrations of fuel and oxygen, the computational grid cannot be fine enough to resolve the combustion reaction, so approximations must be made for the temperature distribution within each grid cell, as well as for the combustion reaction rates themselves.

NOWCASTING METEOROLOGY

Nowcasting comprises the detailed description of the current weather along with forecasts obtained by extrapolation for a period of 0 to 6 hours ahead. In this time range it is possible to forecast small features such as individual storms with reasonable accuracy. A forecaster using the latest radar, satellite and observational data is able to make analysis of the small-scale features present in a small area such as a city and make an accurate forecast for the following few hours. It is, therefore, a powerful tool in warning the public of hazardous, high-impact weather including tropical cyclones, thunderstorms and tornados which cause flash floods, lightning strikes and destructive winds. In broad terms, nowcasting contributes to the:

- Reduction of fatalities and injuries due to weather hazards;
- Reduction of private, public, and industrial, property damage; and to
- Improved efficiency and savings for industry, transportation and agriculture.

In addition to using Nowcasting for warning the public of hazardous weather, it is also used for aviation weather forecasts in both the terminal and en-route environment, marine safety, water and power management, off-shore oil drilling, construction industry and leisure industry. The strength of nowcasting lies in the fact that it provides location-specific forecasts of storm initiation, growth, movement and dissipation, which allows for specific preparation for a certain weather event by people in a specific location.

Extrapolating radar echoes is the mainstay of Nowcasting. This is because radar data is very detailed and picks out the size, shape, intensity, speed and direction of movement of individual storms on a practically continuous basis. This makes it possible to extrapolate the likely location of a moving storm. The intensity of rainfall from a particular cloud or group of clouds can be estimated, which gives a very good indication as to whether to expect flooding, the swelling of a river etc. Depending on the area of built-up space, drainage and land-use in general, a forecast warning tailored to the safety and well-being of a particular community may be issued. In order to extend the time period of Nowcasting beyond 6 hours, some Nowcasting systems use the combination of radar extrapolation techniques with satellite and Numerical Weather Prediction (NWP) model products to produce an extended short-period forecast.

Given the utility of Nowcasting, it is desirable that every National Meteorological and Hydrological Service (NMHS) develops its PWS capacity to effectively render Nowcasting Services to the public.

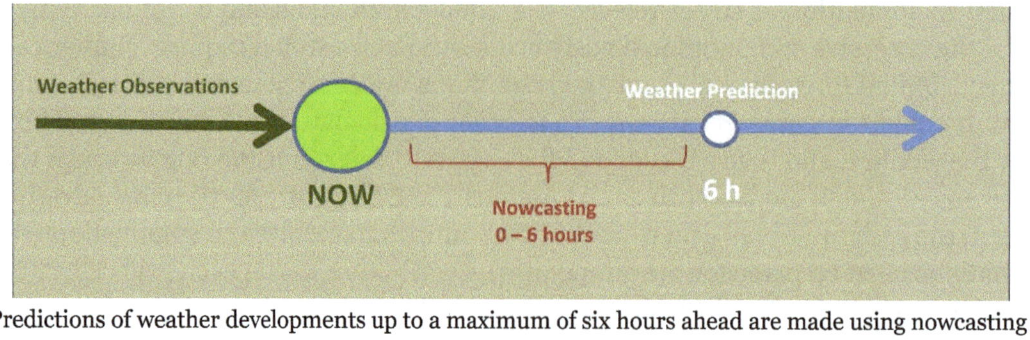

Predictions of weather developments up to a maximum of six hours ahead are made using nowcasting

Nowcasting involves high spatial and temporal resolution forecasts of weather developments for the next few minutes and up to a maximum of six hours ahead. MeteoSwiss uses these short-term forecasts to, among other things, predict thunderstorms, hail and heavy rainfall. This means that nowcasting is an indispensable tool for civil protection.

Short-term forecasts can be generated for any number of meteorological parameters, such as temperature, sunshine or precipitation. The main focus of nowcasting, however, is on forecasting and issuing warnings for phenomena which have particular practical implications, such as severe thunderstorms, heavy rain, hail and snow, as well as wind and fog. The systems developed for this purpose combine all available information in real time in order to predict something as precisely as possible, such as the amount of rain in a river basin in the coming few hours. Measurements and model data form the basis of these short-term forecasts. The measurement data from the weather radar network and meteorological satellites play a particularly important role here.

Applications of Nowcasting

MeteoSwiss uses nowcasting for a wide variety of applications. Many of these help in the early detection of hazards such as thunderstorms or heavy rainfall.

Nowcasting in Civil Protection: the Example of the Thunderstorm

Thunderstorms occur on a regular basis during the six months around summer. This can lead to local flash flooding and landslides, particularly in the Alps and pre-Alpine regions. In addition, strong gusts of wind and large hailstones up to several centimetres in diameter can occur in the vicinity of a thunderstorm. Thunderstorms endanger humans and animals, as well as causing considerable damage every year to vehicles, buildings and agriculture, and disrupting road, rail and air traffic. It is therefore imperative that up-to-date weather information and forecasts for the coming few hours are available in order to allow the relevant authorities, event organisers and individuals to take the necessary precautions in good time. Nowcasting is therefore one of the cornerstones of civil protection.

The weather warnings generated using nowcasting techniques are issued by MeteoSwiss, available i.a. by email or smartphone app. The Swiss government's natural hazards portal provides information on current hazards.

Limitations of Numerical Weather Models

Numerical weather models such as COSMO, used by MeteoSwiss, calculate regional weather forecasts for a few days ahead. From these forecasts, the probability of thunderstorms can also be estimated. However, there are limitations as to how accurately the time and location can be predicted in which heavy precipitation, hail and thunderstorm cells will occur. This is primarily because there is not sufficiently precise information available to the weather models on the temperature and humidity distribution or the state of the clouds at the start of the forecast, in order to be able to generate a forecast that is accurate to within a few kilometres and minutes.

Combination of Numerous Weather Data in Real Time

With the nowcasting systems developed by MeteoSwiss, factors such as the position and strength of thunderstorms, heavy rain and hail cells can be forecast for up to six hours ahead. For this purpose, radar, satellite and lightning measurements in real time are combined with model data to produce as detailed and precise predictions as possible on the development and path of thunderstorm cells. Initial nowcasting systems were restricted to the use of weather radar data, however, the new generation includes additional monitoring data in order to further improve the forecasting. Together with those from radar, measurements from satellites, lightning and meteorological stations, and the forecasts from the COSMO weather model are also used. Algorithms that amalgamate numerous information sources are known as expert systems. These allow the characteristics and the path of thunderstorm cells to be tracked over a period of time. On the basis of changes in the storm cell, estimates can be made as to the potential for hazards within the coming hours.

The expert systems TRT and COALITION developed by MeteoSwiss have now been in operation for several years. The forecasters obtain information from these systems about the current and future developments of thunderstorms, and can thus convey warnings promptly to all the relevant official institutions and the media in the case of a prevailing or impending hazard.

Role of Measurement Data

A key factor in the success of nowcasting is the rapid communication and processing of measurement data from different monitoring systems. The most important of these are data from weather satellites and weather radar, as well as signals from lightning localisation antennae. The combination of these systems allows optimal observation of clouds, thunderstorms, heavy rain cells and hail cells in their various stages of development. The satellite data are particularly useful for observing the initial cloud formation phase when there is no precipitation. Radar data then provide further information on clouds as soon as precipitation has begun. During the most intensive phase of a thunderstorm, all three systems can be used to ascertain the characteristics of the storm.

References

- John D. Cox (2002). Stormwatchers: The Turbulent History of Weather Prediction From Franklin's Kite to El Nino. John Wiley & Sons, Inc. Pp. 53–56. ISBN 978-0-471-38108-2
- Weather-forecasting, science: britannica.com, Retrieved 3 July, 2019
- Datastreme Atmosphere (2008-04-28). "Air Temperature Patterns". American Meteorological Society. Archived from the original on 2008-05-11. Retrieved 2010-02-07
- Surface-weather-analysis : ipfs.io, Retrieved 4 August, 2019
- William J. Emery; Richard E. Thomson (2001). Data analysis methods in physical oceanography. Eos Transactions. 80. Gulf Professional Publishing. Pp. 24–25. Bibcode:1999eostr..80..106J. Doi:10.1029/99EO00074. ISBN 978-0-444-50757-0
- Nowcasting, warning-and-forecasting-systems, measurement-and-forecasting-systems: meteoswiss.admin.ch, Retrieved 5 January, 2019
- Chan, Johnny C. L. & Jeffrey D. Kepert (2010). Global Perspectives on Tropical Cyclones: From Science to Mitigation. World Scientific. Pp. 295–296. ISBN 978-981-4293-47-1. Retrieved 2011-02-24

Meteorological Instruments

5 CHAPTER

Various equipment are used in meteorology such as hygrometer, anemometer, barometer, ceiling balloon, radiosonde, SODAR, barograph, wind profiler, ceiling projector, dark adaptor goggles, windsock, etc. The topics elaborated in this chapter will help in gaining a better perspective about these equipments used in meteorology.

HYGROMETER

A hygrometer is a device used to measure humidity. Many hygrometer products also measure temperature. A hygrometer that measures both humidity and temperature often is referred to as a thermo-hygrometer. PCE Instruments (PCE) offers a wide variety of hygrometer products used to measure the relative humidity of the air.

Since humidity is an important factor in many sectors including but not limited to manufacturing, agriculture, construction, restoration, remediation and heating, ventilation and air conditioning (HVAC), PCE also carries a number of hygrometer devices designed for specific applications such as measuring the humidity in bales of pressed hay or in rolls of paper. A hygrometer also can be used to evaluate heat stress risk in hot work environments like boiler rooms, factories, construction sites and commercial bakeries. To accurately assess heat stress risk, a hygrometer capable of detecting and classifying temperature in relation to the amount of moisture content in the air is recommended. The global standard description for heat stress is the Wet Bulb Globe Temperature Index (WBGT).

A combination hygrometer / thermometer, or thermo-hygrometer, can be used to verify the integrity of the cold chain during the transport and warehousing of refrigerated food and pharmaceutical products. This versatile and popular style of hygrometer also is ideal for evaluating the performance of HVAC systems in industrial, commercial and residential properties. If the humidity indoors is too high, mold can occur. Certain types of mold cause respiratory problems and other negative health effects for people living or working in the surrounding environment. In addition, mold that has permeated into the walls or roof of a structure must be removed for safety reasons. Mold remediation and removal, also called mold damage restoration, can be a costly endeavor, which is why preventative monitoring of the humidity indoors is preferable whenever possible.

Hygrometers can be big and complex, or small and included in a handheld device that measures several key atmospheric conditions. They are currently utilized by a number of different industries and professions, including those involved with weather forecasting, professionals who are responsible for maintaining humidity in museums and residential areas, HVAC professionals and those who participate in outdoor sports and outdoor training, such as members of the military.

Small, handheld devices that include a hygrometer among several other instruments can provide fast, accurate readings of the relative humidity and other important atmospheric readings. These electronically powered devices can measure humidity by using materials that have electrical resistance. The resistance changes based on the amount of moisture that is absorbed. True to the hygrometer definition, the two most-commonly used sensors for hygrometers are called capacitive and resistive. The resistive sensors utilize a membrane made from polymers. This type of sensor registers changes in conductivity when water is absorbed. Capacitive sensors work by measuring changes in how much water there is in the air by utilizing an electrical signal between two plates.

There are two commonly used types of hygrometers: Mechanical hygrometer and wet and dry bulb psychrometer.

Wet and Dry Bulb Psychrometer

This is the easiest way to measure humidity. This type of hygrometer is equipped with two mercury thermometers, where one has a wet bulb and the other has a dry bulb.

Because of the evaporation of water on the wet bulb the temperature will drop and read a lower temperature than what is displayed on the dry bulb. The difference between the two temperature readings equal the amount of relative humidity in the atmosphere.

Mechanical Hygrometer

A mechanical hygrometer requires a little more effort to determining humidity levels in a room.

- This tool was first created in 1783 by a physicist named Horace Benedict de Saussure.
- Mechanical hygrometers work by using an organic material, typically a piece of hair where its behaviors can predict the amount of humidity in the air.
- If you've ever noticed how human hair tends to frizz when there is a lot of moisture in the air or it is very hot outside, then it will be easy for you to understand how this tool works.

For example, the piece of hair is attached to a spring and needle instrument that exposes the hair to humidity. Based on the reaction of the hair, the humidity level can be classified.

Although a wet and dry bulb is more accurate and easy to understand, a mechanical hygrometer is still as effective.

Reset your Hygrometer

Should you need to ever reset your hygrometer, you can do so by using at-home methods:

- In a room with normal, consistent temperature, place your hygrometer in a cup or container filled with salt water on a counter space. Leave it to sit for 10-12 hours.
- After the allotted time, the hygrometer should read a standard relative humidity level of 75 percent.
- This process should be performed at least once a year to ensure your hygrometer is always providing accurate results.

What Makes it Effective?

Hygrometers are the go-to source for measuring humidity. This tool can be used in laboratories, manufacturing sites and storage vicinities. Even meteorologists use hygrometers to report the most accurate amount of relative humidity in the community.

Hygrometers are widely used because they come with hard-to-beat features. Many hygrometers are built with alarms that will alert you when the humidity level in your home is under that 30 percent or over the 50 percent average humidity level.

Hygrometers can serve as a great way to keep you, your family, home and belongings healthy. They can also come with humidistats, which control the operation of your humidifier or dehumidifier.

Different from other Ways of Measuring Humidity

Even though there are many ways to measure humidity, hygrometers prove to be the most accurate.

Plants

- Some people use plants as an indicator of humidity.
- Plants tend to have drastic reactions to weather, so identifying the changes that occur can give you a good idea of how humid it is.
- For example, when humidity is low and the atmosphere is very dry, pine cones will expand their spine openings. This may be similar to how a mechanical hygrometer works, but may not be as precise when measuring specific amounts and percentages such as relative humidity.

Mold

When rooms or areas are overly humid, a stale, mildew odor can form. Even mold can form in some areas. Mold can lead to many health issues and also damage the upholstering of your furniture and belongings.

Peeling Paint

Not enough moisture in your home can cause the paint on the wall to chip off and instruments to warp. All of these problems can be easily avoided with the purchase of a hygrometer.

Proper Measure of Relative Humidity

Not taking the proper precautions and measuring humidity in your home can lead to serious health conditions. This is why it is very important to take advantage of tools used to monitor humidity in your home. If you do not have the tools needed to perform humidity calculations, your first warning of low or above average relative humidity can be a trip to your physician:

- Heat Exhaustion: Heat exhaustion can occur when there is too much humidity. This is common during summer where the temperatures are high. Not drinking enough fluids during humidity peaks is a main cause for heat exhaustion.
- Dehydration: Dehydration can also lead to heat exhaustion. When humidity is high and causes your body to sweat profusely, you are not only losing water supply from your body but also putting more activity on your than it can bear.
- Dry Skin: When humidity levels are low in your home, your skin may become flaky and dry. Some people even experience chapped lips as well.
- Heat Stroke: This is by far the most serious health risk that one could encounter from high humidity levels. Taking all of the described precautions above will help prevent your body from suffering from a heat stroke.

Standard Parts

Because there are different types of hygrometers, there a many components that make up the appliance.

- The psychrometer, as described earlier, has two thermometers attached on each side. It also comes with two bulbs, where one is shielded with muslin.

- A mechanical hygrometer is furnished with a spring and needle device that allow both to work together to pull the hair tightly.
    - Hygrometers such as electrical and chemical hygrometers uses similar tools on substances such as lithium chloride and phosphorus pentoxide to measure humidity.
    - Most parts on hygrometers are fairly simple to understand and assemble.
    - Instructions are usually provided on ways to upkeep the parts for a long life and maximum use.

ANEMOMETER

A hemispherical cup anemometer of the type invented in 1846 by John Thomas Romney Robinson

An anemometer is a device used for measuring wind speed, and is also a common weather station instrument. The term is derived from the Greek word *anemos*, which means wind, and is used to describe any wind speed instrument used in meteorology. The first known description of an anemometer was given by Leon Battista Alberti in 1450.

Velocity Anemometers

Cup Anemometers

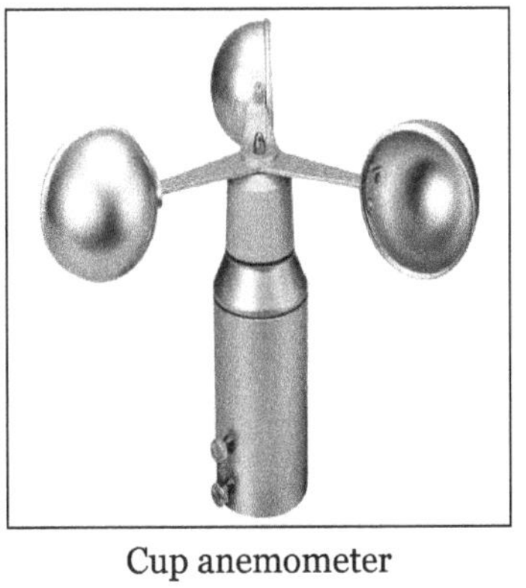

Cup anemometer

A simple type of anemometer was invented in 1845 by Dr. John Thomas Romney Robinson, of Armagh Observatory. It consisted of four hemispherical cups mounted on horizontal arms, which were mounted on a vertical shaft. The air flow past the cups in any horizontal direction turned the shaft at a rate that was roughly proportional to the wind speed. Therefore, counting the turns of the shaft over a set time interval produced a value proportional to the average wind speed for a wide range of speeds. It is also called a rotational anemometer.

On an anemometer with four cups, it is easy to see that since the cups are arranged symmetrically on the end of the arms, the wind always has the hollow of one cup presented to it and is blowing on the back of the cup on the opposite end of the cross. Since a hollow hemisphere has a drag coefficient of .38 on the spherical side and 1.42 on the hollow side, more force is generated on the cup that is presenting its hollow side to the wind. Because of this asymmetrical force, torque is generated on the axis of the anemometer, causing it to spin.

Theoretically, the speed of rotation of the anemometer should be proportional to the wind speed, because the force produced on an object is proportional to the speed of the fluid flowing past it, but other factors influence the rotational speed, including turbulence produced by the apparatus, increasing drag in opposition to the torque that is produced by the cups and support arms, and friction of the mount point. When Robinson first designed his anemometer, he asserted that the cups moved one-third of the speed of the wind, unaffected by the cup size or arm length. This was apparently confirmed by some early independent experiments, but it was incorrect. Instead, the ratio of the speed of the wind and that of the cups, the *anemometer factor*, depends on the dimensions of the cups and arms, and may have a value between two and a little over three. Every previous experiment involving an anemometer had to be repeated after the error was discovered.

The three-cup anemometer developed by the Canadian John Patterson in 1926 and subsequent cup improvements by Brevoort & Joiner of the United States in 1935 led to a cupwheel design with a nearly linear response and had an error of less than 3% up to 60 mph (97 km/h). Patterson found that each cup produced maximum torque when it was at 45° to the wind flow. The three-cup anemometer also had a more constant torque and responded more quickly to gusts than the four-cup anemometer.

The three-cup anemometer was further modified by the Australian Dr. Derek Weston in 1991 to measure both wind direction and wind speed. Weston added a tag to one cup, which causes the cupwheel speed to increase and decrease as the tag moves alternately with and against the wind. Wind direction is calculated from these cyclical changes in cupwheel speed, while wind speed is determined from the average cupwheel speed.

Three-cup anemometers are currently used as the industry standard for wind resource assessment studies & practice.

Vane Anemometers

One of the other forms of mechanical velocity anemometer is the *vane anemometer*. It may be described as a windmill or a propeller anemometer. Unlike the Robinson anemometer, whose axis of rotation is vertical, the vane anemometer must have its axis parallel to the direction of the wind and therefore horizontal. Furthermore, since the wind varies in direction and the axis has to follow its changes, a wind vane or some other contrivance to fulfill the same purpose must be employed.

A *vane anemometer* thus combines a propeller and a tail on the same axis to obtain accurate and precise wind speed and direction measurements from the same instrument. The speed of the fan is measured by a rev counter and converted to a windspeed by an electronic chip. Hence, volumetric flow rate may be calculated if the cross-sectional area is known.

In cases where the direction of the air motion is always the same, as in ventilating shafts of mines and buildings, wind vanes known as air meters are employed, and give satisfactory results.

Helicoid propeller anemometer incorporating a wind vane for orientation

Hand-held digital anemometer

Hot-wire Anemometers

Hot wire anemometers use a fine wire (on the order of several micrometres) electrically heated to some temperature above the ambient. Air flowing past the wire cools the wire. As the electrical resistance of most metals is dependent upon the temperature of the metal (tungsten is a popular choice for hot-wires), a relationship can be obtained between the resistance of the wire and the flow speed.

Several ways of implementing this exist, and hot-wire devices can be further classified as CCA (constant current anemometer), CVA (constant voltage anemometer) and CTA (constant-temperature anemometer). The voltage output from these anemometers is thus the result of some sort of circuit within the device trying to maintain the specific variable (current, voltage or temperature) constant, following Ohm's law.

Additionally, PWM (pulse-width modulation) anemometers are also used, wherein the velocity is inferred by the time length of a repeating pulse of current that brings the wire up to a specified resistance and then stops until a threshold "floor" is reached, at which time the pulse is sent again.

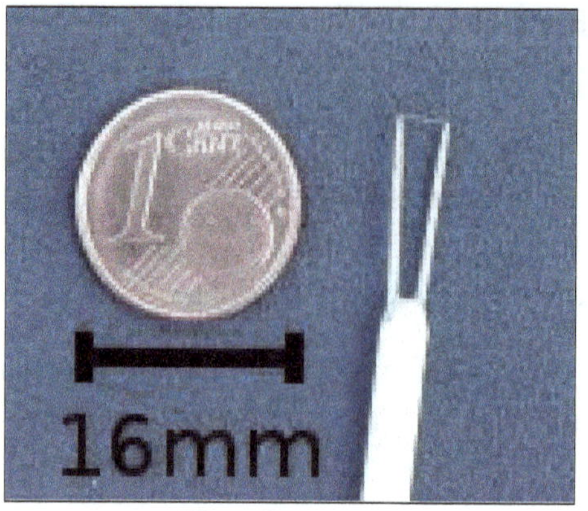

Hot-wire sensor

Hot-wire anemometers, while extremely delicate, have extremely high frequency-response and fine spatial resolution compared to other measurement methods, and as such are almost universally employed for the detailed study of turbulent flows, or any flow in which rapid velocity fluctuations are of interest.

An industrial version of the fine-wire anemometer is the thermal flow meter, which follows the same concept, but uses two pins or strings to monitor the variation in temperature. The strings contain fine wires, but encasing the wires makes them much more durable and capable of accurately measuring air, gas, and emissions flow in pipes, ducts, and stacks. Industrial applications often contain dirt that will damage the classic hot-wire anemometer.

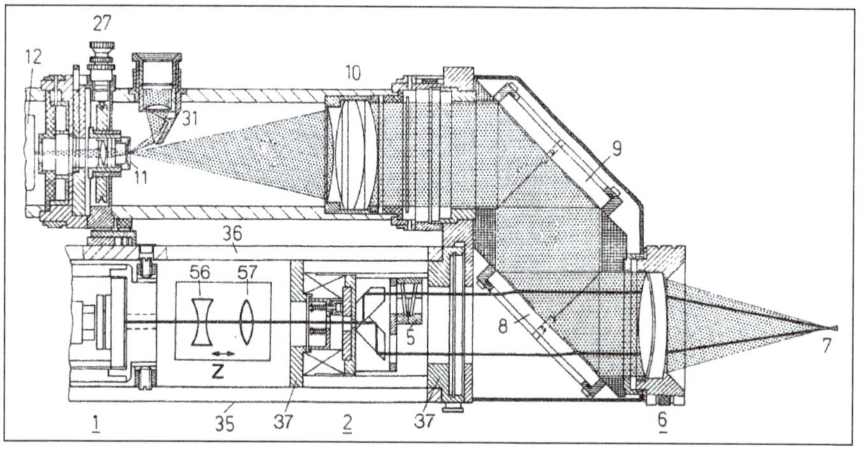

Drawing of a laser anemometer. The laser light is emitted (1) through the front lens (6) of the anemometer and is backscattered off the air molecules (7). The backscattered radiation (dots) re-enters the device and is reflected and directed into a detector (12)

Laser Doppler Anemometers

In laser Doppler velocimetry, laser Doppler anemometers use a beam of light from a laser that is divided into two beams, with one propagated out of the anemometer. Particulates (or deliberately introduced seed material) flowing along with air molecules near where the beam exits reflect, or backscatter, the light back into a detector, where it is measured relative to the original laser beam. When the particles are in great motion, they produce a Doppler shift for measuring wind speed in the laser light, which is used to calculate the speed of the particles, and therefore the air around the anemometer.

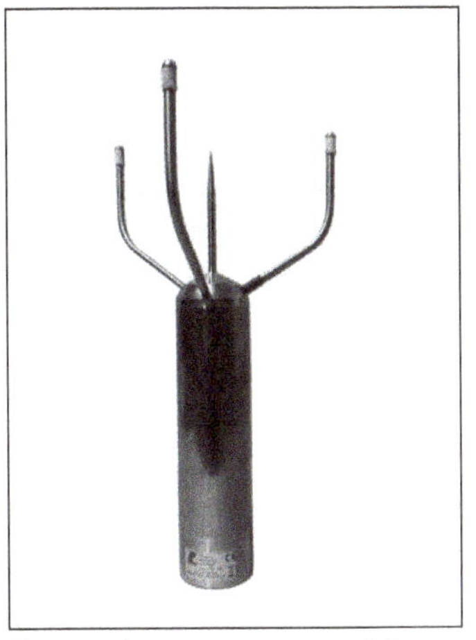

2D ultrasonic anemometer with 3 paths

Ultrasonic Anemometers

Ultrasonic anemometers, first developed in the 1950s, use ultrasonic sound waves to measure wind velocity. They measure wind speed based on the time of flight of sonic pulses between pairs of transducers. Measurements from pairs of transducers can be combined to yield a measurement of velocity in 1-, 2-, or 3-dimensional flow. The spatial resolution is given by the path length between transducers, which is typically 10 to 20 cm. Ultrasonic anemometers can take measurements with very fine temporal resolution, 20 Hz or better, which makes them well suited for turbulence measurements. The lack of moving parts makes them appropriate for long-term use in exposed automated weather stations and weather buoys where the accuracy and reliability of traditional cup-and-vane anemometers are adversely affected by salty air or dust. Their main disadvantage is the distortion of the air flow by the structure supporting the transducers, which requires a correction based upon wind tunnel measurements to minimize the effect. An international standard for this process, ISO 16622 *Meteorology—Ultrasonic anemometers/thermometers—Acceptance test methods for mean wind measurements* is in general circulation. Another disadvantage is lower accuracy due to precipitation, where rain drops may vary the speed of sound.

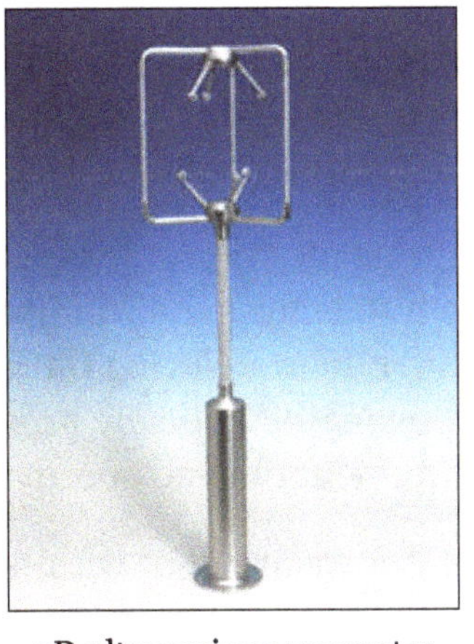

3D ultrasonic anemometer

Since the speed of sound varies with temperature, and is virtually stable with pressure change, ultrasonic anemometers are also used as thermometers.

Two-dimensional (wind speed and wind direction) sonic anemometers are used in applications such as weather stations, ship navigation, aviation, weather buoys and wind turbines. Monitoring wind turbines usually requires a refresh rate of wind speed measurements of 3 Hz, easily achieved by sonic anemometers. Three-dimensional sonic anemometers are widely used to measure gas emissions and ecosystem fluxes using the eddy covariance method when used with fast-response infrared gas analyzers or laser-based analyzers.

Two-dimensional wind sensors are of two types:

- Two ultrasounds paths: These sensors have four arms. The disadvantage of this type of sensor is that when the wind comes in the direction of an ultrasound path, the arms disturb the airflow, reducing the accuracy of the resulting measurement.
- Three ultrasounds paths: These sensors have three arms. They give one path redundancy of the measurement which improves the sensor accuracy and reduces aerodynamic turbulence.

Acoustic Resonance Anemometers

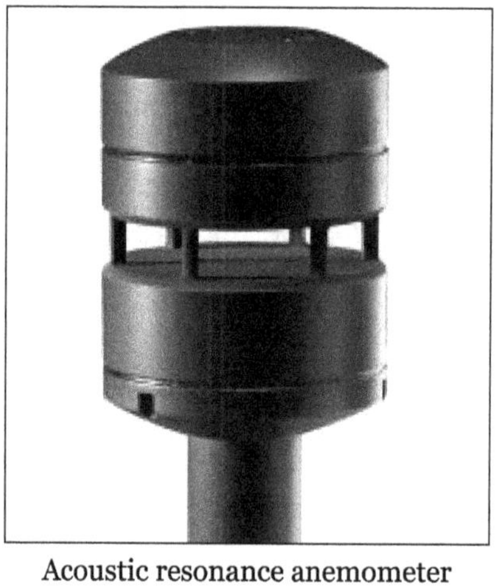

Acoustic resonance anemometer

Acoustic resonance anemometers are a more recent variant of sonic anemometer. The technology was invented by Savvas Kapartis and patented in 1999. Whereas conventional sonic anemometers rely on time of flight measurement, acoustic resonance sensors use resonating acoustic (ultrasonic) waves within a small purpose-built cavity in order to perform their measurement.

Built into the cavity is an array of ultrasonic transducers, which are used to create the separate standing-wave patterns at ultrasonic frequencies. As wind passes through the cavity, a change in the wave's property occurs (phase shift). By measuring the amount of phase shift in the received signals by each transducer, and then by mathematically processing the data, the sensor is able to provide an accurate horizontal measurement of wind speed and direction.

Acoustic resonance technology enables measurement within a small cavity, the sensors therefore tend to be typically smaller in size than other ultrasonic sensors. The small size of acoustic resonance anemometers makes them physically strong and easy to heat and therefore resistant to icing. This combination of features means that they achieve high levels of data availability and are well suited to wind turbine control and to other uses that require small robust sensors such as battlefield meteorology. One issue with this sensor type is measurement accuracy when compared to a calibrated mechanical sensor. For many end uses, this weakness is compensated for by the sensor's longevity and the fact that it does not require re-calibrating once installed.

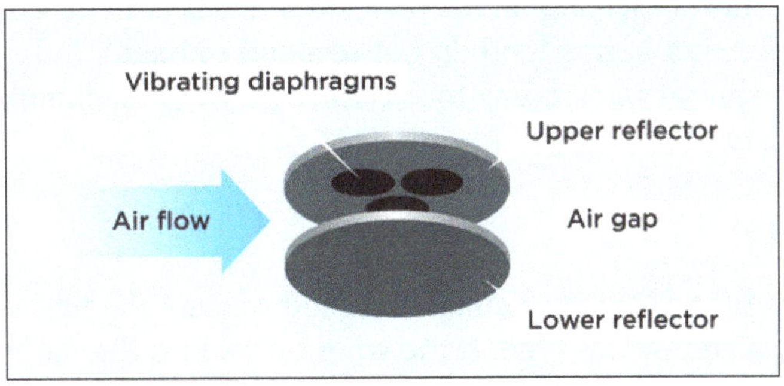

Acoustic resonance principle

Ping-pong Ball Anemometers

A common anemometer for basic use is constructed from a ping-pong ball attached to a string. When the wind blows horizontally, it presses on and moves the ball; because ping-pong balls are very lightweight, they move easily in light winds. Measuring the angle between the string-ball apparatus and the vertical gives an estimate of the wind speed.

This type of anemometer is mostly used for middle-school level instruction, which most students make on their own, but a similar device was also flown on Phoenix Mars Lander.

Pressure Anemometers

The first designs of anemometers that measure the pressure were divided into plate and tube classes.

Plate Anemometers

Britannia Yacht Club clubhouse tour, burgee, and wind gauge on roof

These are the first modern anemometers. They consist of a flat plate suspended from the top so that the wind deflects the plate. In 1450, the Italian art architect Leon Battista Alberti invented the first mechanical anemometer; in 1664 it was re-invented by Robert Hooke (who is often mistakenly considered the inventor of the first anemometer). Later versions of this form consisted of a flat plate, either square or circular, which is kept normal to the wind by a wind vane. The pressure of the wind on its face is balanced by a spring. The compression of the spring determines the actual force which the wind is exerting on the plate, and this is either read off on a suitable gauge, or on a recorder. Instruments of this kind do not respond to light winds, are inaccurate for high wind readings, and are slow at responding to variable winds. Plate anemometers have been used to trigger high wind alarms on bridges.

Tube Anemometers

End remains parallel to the wind flow. Though the Lind was not the first it was the most practical and best known anemometer of this type. If the wind blows into the mouth of a tube it causes an increase of pressure on one side of the manometer. The wind over the open end of a vertical tube causes little change in pressure on the other side of the manometer. The resulting elevation difference in the two legs of the U tube is an indication of the wind speed. However, an accurate measurement requires that the wind speed be directly into the open end of the tube; small departures from the true direction of the wind causes large variations in the reading.

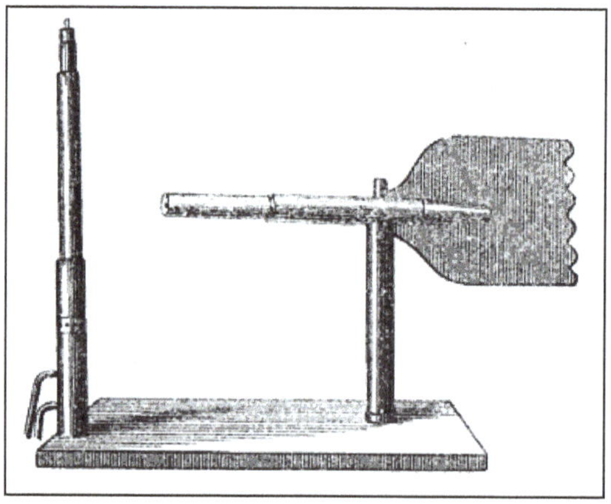

Tube anemometer invented by William Henry Dines. The movable part (right) is mounted on the fixed part (left)

The pointed head is the pitot port. The small holes are connected to the static port

James Lind's anemometer of 1775 consisted of a glass U tube containing a liquid manometer (pressure gauge), with one end bent in a horizontal direction to face the wind and the other vertical.

The successful metal pressure tube anemometer of William Henry Dines in 1892 utilized the same pressure difference between the open mouth of a straight tube facing the wind and a ring of small holes in a vertical tube which is closed at the upper end. Both are mounted at the same height. The pressure differences on which the action depends are very small, and special means are required to register them. The recorder consists of a float in a sealed chamber partially filled with water. The pipe from the straight tube is connected to the top of the sealed chamber and the pipe from the small tubes is directed into the bottom inside the float. Since the pressure difference determines the vertical position of the float this is a measure of the wind speed.

Instruments at Mount Washington Observatory. The pitot tube static anemometer is on the right

The great advantage of the tube anemometer lies in the fact that the exposed part can be mounted on a high pole, and requires no oiling or attention for years; and the registering part can be placed in any convenient position. Two connecting tubes are required. It might appear at first sight as though one connection would serve, but the differences in pressure on which these instruments depend are so minute, that the pressure of the air in the room where the recording part is placed has to be considered. Thus if the instrument depends on the pressure or suction effect alone, and this pressure or suction is measured against the air pressure in an ordinary room, in which the doors and windows are carefully closed and a newspaper is then burnt up the chimney, an effect may be produced equal to a wind of 10 mi/h (16 km/h); and the opening of a window in rough weather, or the opening of a door, may entirely alter the registration.

While the Dines anemometer had an error of only 1% at 10 mph (16 km/h), it did not respond very well to low winds due to the poor response of the flat plate vane required to turn the head into the wind. In 1918 an aerodynamic vane with eight times the torque of the flat plate overcame this problem.

Pitot Tube Static Anemometers

Modern tube anemometers use the same principle as in the Dines anemometer but using a different design. The implementation uses a pitot-static tube which is a pitot tube with two ports, pitot and static, that is normally used in measuring the airspeed of aircraft. The pitot port measures the dynamic pressure of the open mouth of a tube with pointed head facing wind, and the static port measures the static pressure from small holes along the side on that tube. The pitot tube is connected to a tail so that it always makes the tube's head to face the wind. Additionally, the tube is heated to prevent rime ice formation on the tube. There are two lines from the tube down to the devices to measure the difference in pressure of the two lines. The measurement devices can be manometers, pressure transducers, or analog chart recorders.

Effect of Density on Measurements

In the tube anemometer the dynamic pressure is actually being measured, although the scale is usually graduated as a velocity scale. If the actual air density differs from the calibration value, due to differing temperature, elevation or barometric pressure, a correction is required to obtain the actual wind speed. Approximately 1.5% (1.6% above 6,000 feet) should be added to the velocity recorded by a tube anemometer for each 1000 ft (5% for each kilometer) above sea-level.

Effect of Icing

At airports, it is essential to have accurate wind data under all conditions, including freezing precipitation. Anemometry is also required in monitoring and controlling the operation of wind turbines, which in cold environments are prone to in-cloud icing. Icing alters the aerodynamics of an anemometer and may entirely block it from operating. Therefore, anemometers used in these applications must be internally heated. Both cup anemometers and sonic anemometers are presently available with heated versions.

Instrument Location

In order for wind speeds to be comparable from location to location, the effect of the terrain needs to be considered, especially in regard to height. Other considerations are the presence of trees, and both natural canyons and artificial canyons (urban buildings). The standard anemometer height in open rural terrain is 10 meters.

BAROMETER

A barometer is a scientific instrument that is used to measure air pressure in a certain environment. Pressure tendency can forecast short term changes in the weather. Many measurements of air pressure are used within surface weather analysis to help find surface troughs, pressure systems and frontal boundaries.

Barometers and pressure altimeters (the most basic and common type of altimeter) are essentially the same instrument, but used for different purposes. An altimeter is intended to be used at different levels matching the corresponding atmospheric pressure to the altitude, while a barometer is kept at the same level and measures subtle pressure changes caused by weather and elements of weather.

Although Evangelista Torricelli is universally credited with inventing the barometer in 1643, historical documentation also suggests Gasparo Berti, an Italian mathematician and astronomer, unintentionally built a water barometer sometime between 1640 and 1643. French scientist and philosopher René Descartes described the design of an experiment to determine atmospheric pressure as early as 1631, but there is no evidence that he built a working barometer at that time.

On July 27, 1630, Giovanni Battista Baliani wrote a letter to Galileo Galilei explaining an experiment he had made in which a siphon, led over a hill about twenty-one meters high, failed to work. Galileo responded with an explanation of the phenomenon: he proposed that it was the power of a vacuum that held the water up, and at a certain height the amount of water simply became too much and the force could not hold any more, like a cord that can support only so much weight. This was a restatement of the theory of *horror vacui* ("nature abhors a vacuum"), which dates to Aristotle, and which Galileo restated as *resistenza del vacuo*.

Galileo's ideas reached Rome in December 1638 in his *Discorsi*. Raffaele Magiotti and Gasparo Berti were excited by these ideas, and decided to seek a better way to attempt to produce a vacuum other than with a siphon. Magiotti devised such an experiment, and sometime between 1639 and 1641, Berti (with Magiotti, Athanasius Kircher and Niccolò Zucchi present) carried it out.

Four accounts of Berti's experiment exist, but a simple model of his experiment consisted of filling with water a long tube that had both ends plugged, then standing the tube in a basin already full of water. The bottom end of the tube was opened, and water that had been inside of it poured out into the basin. However, only part of the water in the tube flowed out, and the level of the water inside the tube stayed at an exact level, which happened to be 10.3 m (34 ft), the same height Baliani and Galileo had observed that was limited by the siphon. What was most important about this experiment was that the lowering water had left a space above it in the tube which had no intermediate contact with air to fill it up. This seemed to suggest the possibility of a vacuum existing in the space above the water.

Torricelli, a friend and student of Galileo, interpreted the results of the experiments in a novel way. He proposed that the weight of the atmosphere, not an attracting force of the vacuum, held the water in the tube. In a letter to Michelangelo Ricci in 1644 concerning the experiments, he wrote:

> Many have said that a vacuum does not exist, others that it does exist in spite of the repugnance of nature and with difficulty; No one has said that it exists without difficulty and without a resistance from nature. Thus if there can be found a manifest cause from which the resistance can be derived which is felt if we try to make a vacuum, it seems foolish to try to attribute to vacuum those operations which follow evidently from some other cause; and so by making some very easy calculations, the cause assigned (that is, the weight of the atmosphere) ought by itself alone to offer a greater resistance than it does when we try to produce a vacuum.

It was traditionally thought (especially by the Aristotelians) that the air did not have weight: that is, that the kilometers of air above the surface did not exert any weight on the bodies below it. Even Galileo had accepted the weightlessness of air as a simple truth. Torricelli questioned that assumption, and instead proposed that air had weight and that it was the latter (not the attracting force of the vacuum) which held (or rather, pushed) up the column of water. He thought that the level the water stayed at (c. 10.3 m) was reflective of the force of the air's weight pushing on it (specifically, pushing on the water in the basin and thus limiting how much water can fall from the tube into it). In other words, he viewed the barometer as a balance, an instrument for measurement (as opposed to merely being an instrument to create a vacuum), and because he was the first to view it this way, he is traditionally considered the inventor of the barometer (in the sense in which we now use the term).

Because of rumors circulating in Torricelli's gossipy Italian neighborhood, which included that he was engaged in some form of sorcery or witchcraft, Torricelli realized he had to keep his experiment secret to avoid the risk of being arrested. He needed to use a liquid that was heavier than water, and from his previous association and suggestions by Galileo, he deduced by using mercury, a shorter tube could be used. With mercury, which is about 14 times denser than water, a tube only 80 cm was now needed, not 10.5 m.

In 1646, Blaise Pascal along with Pierre Petit, had repeated and perfected Torricelli's experiment after hearing about it from Marin Mersenne, who himself had been shown the experiment by Torricelli toward the end of 1644. Pascal further devised an experiment to test the Aristotelian proposition that it was vapors from the liquid that filled the space in a barometer. His experiment compared water with wine, and since the latter was considered more "spiritous", the Aristotelians expected the wine to stand lower (since more vapors would mean more pushing down on the liquid column). Pascal performed the experiment publicly, inviting the Aristotelians to predict the outcome beforehand. The Aristotelians predicted the wine would stand lower. It did not.

However, Pascal went even further to test the mechanical theory. If, as suspected by mechanical philosophers like Torricelli and Pascal, air had weight, the pressure would be less at higher altitudes. Therefore, Pascal wrote to his brother-in-law, Florin Perier, who lived near a mountain called the Puy de Dome, asking him to perform a crucial experiment. Perier was to take a barometer up the Puy de Dome and make measurements along the way of the height of the column of mercury. He was then to compare it to measurements taken at the foot of the mountain to see if those measurements taken higher up were in fact smaller. In September 1648, Perier carefully and meticulously carried out the experiment, and found that Pascal's predictions had been correct. The mercury barometer stood lower the higher one went.

Types

Water-based Barometers

Goethe's device

The concept that decreasing atmospheric pressure predicts stormy weather, postulated by Lucien Vidi, provides the theoretical basis for a weather prediction device called a "weather glass" or a "Goethe barometer" (named for Johann Wolfgang Von Goethe, the renowned German writer and polymath who developed a simple but effective weather ball barometer using the principles developed by Torricelli). The French name, *le baromètre Liègeois,* is used by some English speakers. This name reflects the origins of many early weather glasses – the glass blowers of Liège, Belgium.

The weather ball barometer consists of a glass container with a sealed body, half filled with water. A narrow spout connects to the body below the water level and rises above the water level. The narrow spout is open to the atmosphere. When the air pressure is lower than it was at the time the body was sealed, the water level in the spout will rise above the water level in the body; when the air pressure is higher, the water level in the spout will drop below the water level in the body. A variation of this type of barometer can be easily made at home.

Mercury Barometers

A mercury barometer has a vertical glass tube closed at the top sitting in an open mercury-filled basin at the bottom. Mercury in the tube adjusts until the weight of the mercury column balances the atmospheric force exerted on the reservoir. High atmospheric pressure places more force on the reservoir, forcing mercury higher in the column. Low pressure allows the mercury to drop to a lower level in the column by lowering the force placed on the reservoir. Since higher temperature levels around the instrument will reduce the density of the mercury, the scale for reading the height of the mercury is adjusted to compensate for this effect. The tube has to be at least as long as the amount dipping in the mercury + head space + the maximum length of the column.

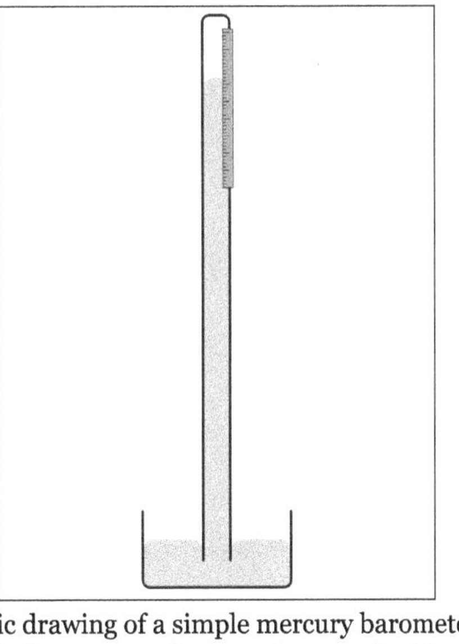

Schematic drawing of a simple mercury barometer with vertical mercury column and reservoir at base

Torricelli documented that the height of the mercury in a barometer changed slightly each day and concluded that this was due to the changing pressure in the atmosphere. He wrote: "We live submerged at the bottom of an ocean of elementary air, which is known by incontestable experiments to have weight". Inspired by Torricelli, Otto von Guericke on 5 December 1660 found that air pressure was unusually low and predicted a storm, which occurred the next day.

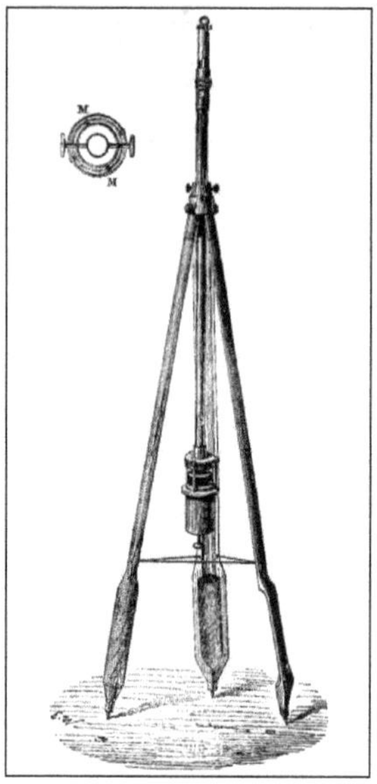

Fortin barometer

The mercury barometer's design gives rise to the expression of atmospheric pressure in inches or millimeters of mercury (mmHg). A torr was originally defined as 1 mmHg. The pressure is quoted as the level of the mercury's height in the vertical column. Typically, atmospheric pressure is measured between 26.5 inches (670 mm) and 31.5 inches (800 mm) of Hg. One atmosphere (1 atm) is equivalent to 29.92 inches (760 mm) of mercury.

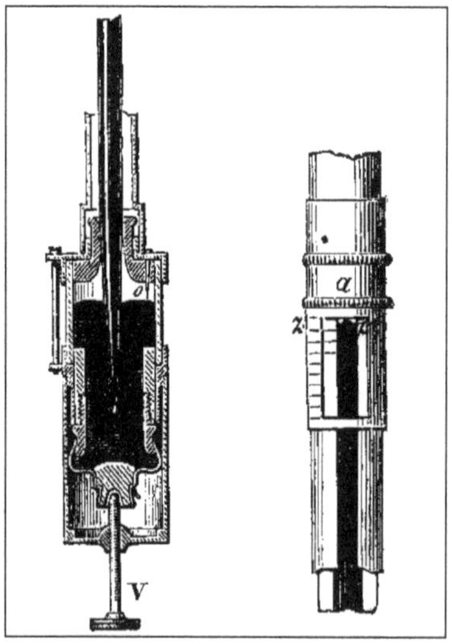

Reservoir of a Fortin barometer

Design changes to make the instrument more sensitive, simpler to read, and easier to transport resulted in variations such as the basin, siphon, wheel, cistern, Fortin, multiple folded, stereometric, and balance barometers. *Fitzroy* barometers combine the standard mercury barometer with a thermometer, as well as a guide of how to interpret pressure changes. Fortin barometers use a variable displacement mercury cistern, usually constructed with a thumbscrew pressing on a leather diaphragm bottom (V in the diagram). This compensates for displacement of mercury in the column with varying pressure. To use a Fortin barometer, the level of mercury is set to zero by using the thumbscrew to make an ivory pointer (O in the diagram) just touch the surface of the mercury. The pressure is then read on the column by adjusting the vernier scale so that the mercury just touches the sightline at Z.. Some models also employ a valve for closing the cistern, enabling the mercury column to be forced to the top of the column for transport. This prevents water-hammer damage to the column in transit.

On June 2007, a European Union directive was enacted to restrict the sale of mercury, thus effectively ending the production of new mercury barometers in Europe.

Vacuum Pump Oil Barometer

Using vacuum pump oil as the working fluid in a barometer has led to the creation of the new "World's Tallest Barometer" in February 2013. The barometer at Portland State University (PSU) uses doubly distilled vacuum pump oil and has a nominal height of about 12.4 m for the oil column height; expected excursions are in the range of ±0.4 m over the course of a year. Vacuum pump oil has very low vapor pressure and it is available in a range of densities; the lowest density vacuum oil was chosen for the PSU barometer to maximize the oil column height.

Aneroid Barometers

Old aneroid barometer

Modern aneroid barometer

An aneroid barometer is an instrument used for measuring pressure as a method that does not involve liquid. Invented in 1844 by French scientist Lucien Vidi, the aneroid barometer uses a small, flexible metal box called an aneroid cell (capsule), which is made from an alloy of beryllium and copper. The evacuated capsule (or usually several capsules, stacked to add up their movements) is prevented from collapsing by a strong spring. Small changes in external air pressure cause the cell to expand or contract. This expansion and contraction drives mechanical levers such that the tiny movements of the capsule are amplified and displayed on the face of the aneroid barometer. Many models include a manually set needle which is used to mark the current measurement so a change can be seen. This type of barometer is common in homes and in recreational boats. It is also used in meteorology, mostly in barographs and as a pressure instrument in radiosondes.

Barographs

A barograph is a recording aneroid barometer where the changes in atmospheric pressure are recorded on a paper chart.

Principle and Working

The principle of the barograph is same as that of the aneroid baromoter. A metal box partially exhausted of air will undergo changes of shape as the outside pressure varies. These small movements of the box are transmitted by a system of levers to a recording arm that has at its extreme end either a scribe or a pen. A scribe records on smoked foil while a pen records on paper using ink, held in a nib. The recording material is mounted on a cylindrical drum which is rotated slowly by a clock. Commonly, the drum makes one revolution per day, per week, or per month and the rotation rate can often be selected by the user.

MEMS Barometers

Micro Electro Mechanical Systems (or MEMS) barometers are extremely small devices between 1 and 100 micrometres in size (0.001 to 0.1 mm). They are created via photolithography or photochemical machining. Typical applications include miniaturized weather stations, electronic barometers and altimeters.

A barometer can also be found in smartphones such as the Samsung Galaxy Nexus, Samsung Galaxy S3-S6, Motorola Xoom, Apple iPhone 6 smartphones, and Timex Expedition WS4 smartwatch, based on MEMS and piezoresistive pressure-sensing technologies. Inclusion of barometers on smartphones was originally intended to provide a faster GPS lock. However, third party researchers were unable to confirm additional GPS accuracy or lock speed due to barometric readings. The researchers suggest that the inclusion of barometers in smartphones may provide a solution for determining a user's elevation, but also suggest that several pitfalls must first be overcome.

More Unusual Barometers

There are many other more unusual types of barometer. From variations on the storm barometer, such as the Collins Patent Table Barometer, to more traditional-looking designs such as Hooke's Otheometer and the Ross Sympiesometer. Some, such as the Shark Oil barometer, work only in a certain temperature range, achieved in warmer climates.

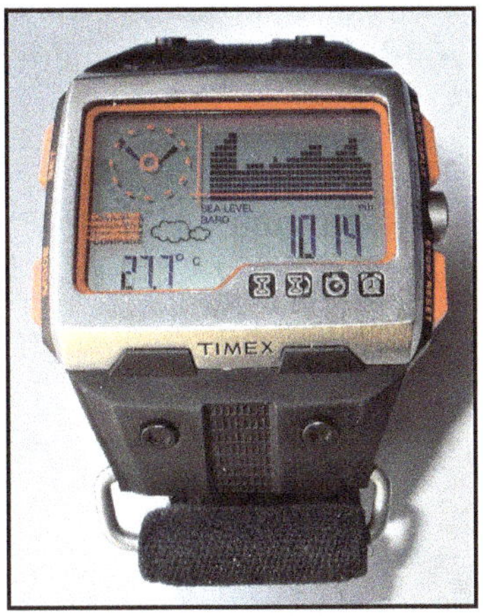

Timex Expedition WS4 in Barometric chart mode with weather forecast function

The Galaxy Nexus has a built-in barometer

Applications

Barometric pressure and the pressure tendency (the change of pressure over time) have been used in weather forecasting since the late 19th century. When used in combination with wind observations, reasonably accurate short-term forecasts can be made. Simultaneous barometric readings from across a network of weather stations allow maps of air pressure to be produced, which were the first form of the modern weather map when created in the 19th century. Isobars, lines of equal pressure, when drawn on such a map, give a contour map showing areas of high and low pressure. Localized high atmospheric pressure acts as a barrier to approaching weather systems, diverting their course. Atmospheric lift caused by low-level wind convergence into the surface brings clouds and sometimes precipitation. The larger the change in pressure, especially if more than 3.5 hPa (0.1 inHg), the greater the change in weather that can be expected. If the pressure drop is rapid, a low pressure system is approaching, and there is a greater chance of rain. Rapid pressure rises, such as in the wake of a cold front, are associated with improving weather conditions, such as clearing skies.

Digital graphing barometer

With falling air pressure, gases trapped within the coal in deep mines can escape more freely. Thus low pressure increases the risk of firedamp accumulating. Collieries therefore keep track of the pressure. In the case of the Trimdon Grange colliery disaster of 1882 the mines inspector drew attention to the records and in the report stated "the conditions of atmosphere and temperature may be taken to have reached a dangerous point".

Analogue recording Barograph using five stacked aneroid barometer cells

Aneroid barometers are used in scuba diving. A submersible pressure gauge is used to keep track of the contents of the diver's air tank. Another gauge is used to measure the hydrostatic pressure, usually expressed as a depth of sea water. Either or both gauges may be replaced with electronic variants or a dive computer.

Compensations

Temperature

The density of mercury will change with increase or decrease in temperature, so a reading must be adjusted for the temperature of the instrument. For this purpose a mercury thermometer is usually mounted on the instrument. Temperature compensation of an aneroid barometer is accomplished by including a bi-metal element in the mechanical linkages. Aneroid barometers sold for domestic use typically have no compensation under the assumption that they will be used within a controlled room temperature range.

Altitude

A digital barometer with altimeter setting (for correction) displayed

As the air pressure decreases at altitudes above sea level (and increases below sea level) the uncorrected reading of the barometer will depend on its location. The reading is then adjusted to an equivalent sea-level pressure for purposes of reporting. For example, if a barometer located at sea level and under fair weather conditions is moved to an altitude of 1,000 feet (305 m), about 1 inch of mercury (~35 hPa) must be added on to the reading. The barometer readings at the two locations should be the same if there are negligible changes in time, horizontal distance, and temperature. If this were not done, there would be a false indication of an approaching storm at the higher elevation.

Aneroid barometers have a mechanical adjustment that allows the equivalent sea level pressure to be read directly and without further adjustment if the instrument is not moved to a different altitude. Setting an aneroid barometer is similar to resetting an analog clock that is not at the correct time. Its dial is rotated so that the current atmospheric pressure from a known accurate and nearby barometer (such as the local weather station) is displayed. No calculation is needed, as the source barometer reading has already been converted to equivalent sea-level pressure, and this is transferred to the barometer being set—regardless of its altitude. Though somewhat rare, a few aneroid barometers intended for monitoring the weather are calibrated to manually adjust for altitude. In this case, knowing *either* the altitude or the current atmospheric pressure would be sufficient for future accurate readings.

The table below shows examples for three locations in the city of San Francisco, California. Note the corrected barometer readings are identical, and based on equivalent sea-level pressure. (Assume a temperature of 15 °C.)

Location	Altitude (feet)	Uncorrected P_{atm} (inches Hg)	Corrected P_{atm} (inches Hg)	Altitude (metres)	Uncorrected P_{atm} (hPa)	Corrected P_{atm} (hPa)
City Marina	Sea Level (0)	29.92	29.92	0 m	1013 hPa	1013 hPa
Nob Hill	348	29.55	29.92	106 m	1001 hPa	1013 hPa
Mt. Davidson	928	28.94	29.92	283 m	980 hPa	1013 hPa

In 1787, during a scientific expedition on Mont Blanc, De Saussure undertook research and executed physical experiments on the boiling point of water at different heights. He calculated the height at each of his experiments by measuring how long it took an alcohol burner to boil an amount of water, and by these means he determined the height of the mountain to be 4775 metres. (This later turned out to be 32 metres less than the actual height of 4807 metres). For these experiments De Saussure brought specific scientific equipment, such as a barometer and thermometer. His calculated boiling temperature of water at the top of the mountain was fairly accurate, only off by 0.1 Kelvin.

Based on his findings, the altimeter could be developed as a specific application of the barometer. In the mid-19th century, this method was used by explorers.

When atmospheric pressure is measured by a barometer, the pressure is also referred to as the "barometric pressure". Assume a barometer with a cross-sectional area A, a height h, filled with mercury from the bottom at Point B to the top at Point C. The pressure at the bottom of the barometer, Point B, is equal to the atmospheric pressure. The pressure at the very top, Point C, can be taken as zero because there is only mercury vapor above this point and its pressure is very low relative to the atmospheric pressure. Therefore, one can find the atmospheric pressure using the barometer and this equation:

$$P_{atm} = \rho g h$$

where ρ is the density of mercury, g is the gravitational acceleration, and h is the height of the mercury column above the free surface area. The physical dimensions (length of tube and cross-sectional area of the tube) of the barometer itself have no effect on the height of the fluid column in the tube.

In thermodynamic calculations, a commonly used pressure unit is the "standard atmosphere". This is the pressure resulting from a column of mercury of 760 mm in height at 0 °C. For the density of mercury, use ρ_{Hg} = 13,595 kg/m³ and for gravitational acceleration use g = 9.807 m/s².

If water were used (instead of mercury) to meet the standard atmospheric pressure, a water column of roughly 10.3 m (33.8 ft) would be needed.

Standard atmospheric pressure as a function of elevation:

1 torr = 133.3 Pa = 0.03937 In Hg.

P_{atm}	Altitude	P_{atm}	Altitude
101.325 kPa	Sea Level (0m)	29.92 In Hg	Sea Level (0 ft)
97.71 kPa	305 m	28.86 In Hg	1,000 ft
94.21 kPa	610 m	27.82 In Hg	2,000 ft
89.88 kPa	1,000 m	26.55 In Hg	3,281 ft
84.31 kPa	1,524 m	24.90 In Hg	5,000 ft
79.50 kPa	2,000 m	23.48 In Hg	6,562 ft
69.68 kPa	3,048 m	20.58 In Hg	10,000 ft
54.05 kPa	5,000 m	15.96 In Hg	16,404 ft
46.56 kPa	6,096 m	13.75 In Hg	20,000 ft
37.65 kPa	7,620 m	11.12 In Hg	25,000 ft
32.77 kPa	8,848 m	9.68 In Hg	29,029 ft
26.44 kPa	10,000 m	7.81 In Hg	32,808 ft
11.65 kPa	15,240 m	3.44 In Hg	50,000 ft
5.53 kPa	20,000 m	1.63 In Hg	65,617 ft

CEILING BALLOON

A ceiling balloon also called a pilot balloon or pibal, is used by meteorologists to determine the height of the base of clouds above ground level during daylight hours. In the past, and sometimes today, a theodolite was used to track the balloon in order to determine the speed and direction of winds aloft. The principle behind the ceiling balloon is that timing of a balloon with a known ascent rate (how fast it climbs) from its release until it disappears into the clouds can be used to calculate the height of the bottom of the clouds.

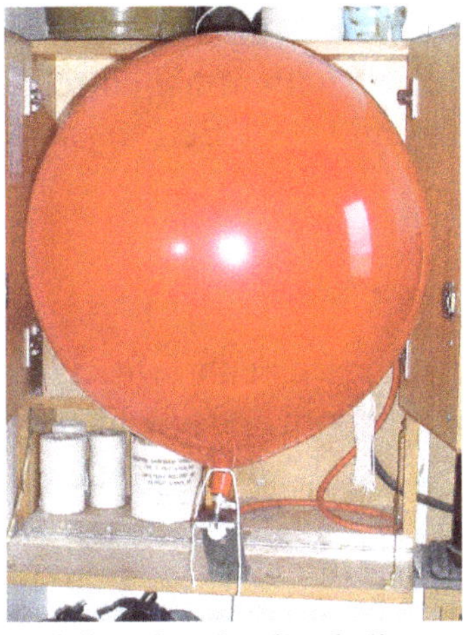

A fully inflated ceiling balloon

Regulator valve and pressure gauges attached to helium cylinder

Use of Ceiling Balloon

A ceiling balloon is a small, usually red, (fluted) rubber balloon commonly measuring 76 mm (3 in) across prior to inflation, inflated to ~40 cm (~15.75 in) diameter. After inflation the balloon is taken outside and released. By timing the balloon from release until it enters the cloud a ceiling height can be obtained. When correctly inflated the balloon will rise at rate of 140 m/min (460 ft/min). The bases of clouds are very rarely flat and solid, so the ceiling height is not when the balloon disappears but when the colour begins to fade. The balloon can also be used to measure the vertical visibility into a layer of fog or blowing snow. In this case the balloon will begin to fade as soon as it is released, so the vertical visibility is when the balloon disappears. If the balloon is visible for a considerable distance into the cloud layer the observer should make note of it as it is of importance to aircraft.

The ceiling balloon is a reliable, safe and simple way to get an indication of the height of clouds. However, it does suffer from some disadvantages that the observer must be aware of. Rain and wet snow may slow the ascent of the balloon, giving a falsely high ceiling and high winds and poor visibility may cause the balloon to appear to enter the cloud before it actually does. As the balloon rises at a rate of 140 m/min (460 ft/min) it will take over five minutes for the balloon to reach 700 m (2300 ft). Beyond this height the ability to follow the balloon, even with binoculars, is poor, as even the slightest movement of the eye off the balloon will almost certainly ensure that it vanishes.

At night when it is not practical to use a balloon the ceiling projector is used. However, during twilight it may be impossible to use the ceiling projector and then a pibal (pilot balloon) light may be used. This is a simple flashlight bulb attached to a battery. To charge the battery it is immersed in water for three minutes and then tied to the balloon prior to inflation. These are rarely used today.

Technical Details

The balloons and associated equipment are usually stored in a cabinet mounted on a wall close to the gas cylinders. The cabinet has three doors one of which opens down and to it the filler stand is attached. At the top of the filler stand is a "L" shaped pipe with two rings, a small one on the bottom and a larger one on the top called the inflation nozzle. The rings stop the tube from dropping through the stand or rising too far when the balloon is inflated. The top ring has several grooves cut into it, to help grip the balloon which is fitted to it.

At the bottom of the pipe is a weight that, when the precise amount of gas has been added, will lift to indicate the balloon is full. A rubber hose is attached to this pipe and passes through the filler stand twice. The first hole is larger than the tube to permit movement, while the second is used to hold the tubing in place.

From there the tube runs to a needle valve that controls the amount of gas flowing to the balloon. A second tube will then run from the valve to a regulator valve that is attached to the gas cylinder. This valve has two pressure gauges attached. One showing the total pressure remaining in the gas cylinder and the second showing the amount of gas flowing through the tubing. Typically the cylinder, which is made of steel and weighs about 140 lb (65 kg). It contains the equivalent of about 200 ft^3 (5.7 m^3) of gas at standard pressure, stored at a pressure of 2000 psi (14 megapascals) and will inflate approximately 120 balloons (according to a nominal diameter of 45 cm).

On the opposite side of the cabinet is space to store balloons, string and pibal lights. The gas used to fill the balloon is helium or hydrogen. Because of its low cost ceiling balloons are often filled with hydrogen gas, but sometimes helium is used.

The balloon is attached to the inflation nozzle and a piece of string is wound around the neck. After donning safety glasses and hearing protection a check is made to ensure the needle valve is fully closed. The main valve on the cylinder is then opened, followed by the regulator valve. Next, the needle valve is opened and the balloon begins to inflate. As the balloon reaches the correct size the inflation nozzle will begin to lift. At this point the needle valve is closed along with the regulator valve and cylinder valve. The string is then used to tie off the balloon neck to ensure that no gas can escape.

Caution must be used during inflation due to the occasional faulty balloon and its failure. If the person inflating the balloon is not wearing goggles or hearing protectors then eye or ear damage can result.

RADIOSONDE

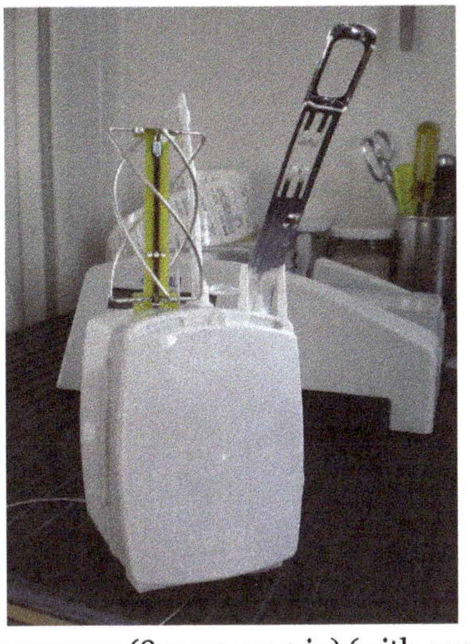

A GPS sonde, approx 220 × 80 ×75 mm (8.7 × 3.1 × 3 in) (with grounding station in background, used to perform a 'ground check' and also recondition the humidity sensor)

A radiosonde is a battery-powered telemetry instrument carried into the atmosphere usually by a weather balloon that measures various atmospheric parameters and transmits them by radio to a ground receiver. Modern radiosondes measure or calculate the following variables: altitude, pressure, temperature, relative humidity, wind (both wind speed and wind direction), cosmic ray readings at high altitude and geographical position (latitude/longitude). Radiosondes measuring ozone concentration are known as ozonesondes.

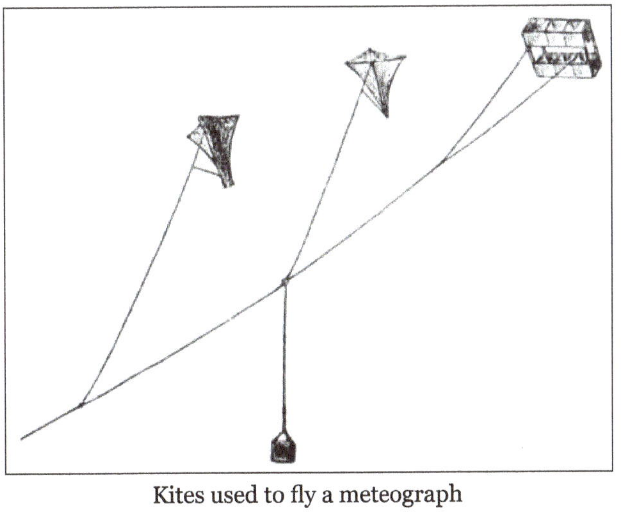

Kites used to fly a meteograph

Radiosondes may operate at a radio frequency of 403 MHz or 1680 MHz. A radiosonde whose position is tracked as it ascends to give wind speed and direction information is called a rawinsonde ("radar wind-sonde"). Most radiosondes have radar reflectors and are technically rawinsondes. A radiosonde that is dropped from an airplane and falls, rather than being carried by a balloon is called a *dropsonde*. Radiosondes are an essential source of meteorological data, and hundreds are launched all over the world daily.

Meteograph used by the US Weather Bureau in 1898

The first flights of aerological instruments were done in the second half of the 19th century with kites and meteographs, a recording device measuring pressure and temperature that was recuperated after the experiment. This proved to be difficult because the kites were linked to the ground and were very difficult to manoeuvre in gusty conditions. Furthermore, the sounding was limited to low altitudes because of the link to the ground.

U.S. Bureau of Standards personnel launch radiosonde near Washington, DC in 1936

Gustave Hermite and Georges Besançon, from France, were the first in 1892 to use a balloon to fly the meteograph. In 1898, Léon Teisserenc de Bort organized at the *Observatoire de Météorologie Dynamique de Trappes* the first regular daily use of these balloons. Data from these launches showed that the temperature lowered with height up to a certain altitude, which varied with the season, and then stabilized above this altitude. De Bort's discovery of the tropopause and stratosphere was announced in 1902 at the French Academy of Sciences. Other researchers, like Richard Aßmann and William Henry Dines, were working at the same times with similar instruments.

US sailors launching a radiosonde during World War 2

In 1924, Colonel William Blaire in the U.S. Signal Corps did the first primitive experiments with weather measurements from balloon, making use of the temperature dependence of radio circuits. The first true radiosonde that sent precise encoded telemetry from weather sensors was invented in France by Robert Bureau. Bureau coined the name "radiosonde" and flew the first instrument on January 7, 1929. Developed independently a year later, Pavel Molchanov flew a radiosonde on January 30, 1930. Molchanov's design became a popular standard because of its simplicity and because it converted sensor readings to Morse code, making it easy to use without special equipment or training.

Working with a modified Molchanov sonde, Sergey Vernov was the first to use radiosondes to perform cosmic ray readings at high altitude. On April 1, 1935, he took measurements up to 13.6 km (8.5 mi) using a pair of Geiger counters in an anti-coincidence circuit to avoid counting secondary ray showers. This became an important technique in the field, and Vernov flew his radiosondes on land and sea over the next few years, measuring the radiation's latitude dependence caused by the Earth's magnetic field.

In 1936, the U.S. Navy assigned the U.S. Bureau of Standards (NBS) to develop an official radiosonde for the Navy to use. The NBS gave the project to Harry Diamond, who had previously worked on radio navigation and invented a blind landing system for airplanes. The organization led by Diamond eventually (in 1992) became a part of the U.S. Army Research Laboratory. In 1937, Diamond, along with his associates Francis Dunmore and Wilbur Hinmann, Jr., created a radiosonde that employed audio-frequency subcarrier modulation with the help of a resistance-capacity relaxation oscillator. In addition, this NBS radiosonde was capable of measuring temperature and humidity at higher altitudes than conventional radiosondes at the time due to the use of electric sensors.

In 1938, Diamond developed the first ground receiver for the radiosonde, which prompted the first service use of the NBS radiosondes in the Navy. Then in 1939, Diamond and his colleagues developed a ground-based radiosonde called the "remote weather station," which allowed them to automatically collect weather data in remote and inhospitable locations. By 1940, the NBS radiosonde system included a pressure drive, which measured temperature and humidity as functions of pressure. It also gathered data on cloud thickness and light intensity in the atmosphere. Due to this and other improvements in cost (about $25), weight (> 1 kilogram), and accuracy, hundreds of thousands of NBS-style radiosondes were produced nationwide for research purposes, and the apparatus was officially adopted by the U.S. Weather Bureau.

Diamond was given the Washington Academy of Sciences Engineering Award in 1940 and the IRE Fellow Award (which was later renamed the Harry Diamond Memorial Award) in 1943 for his contributions to radio-meteorology.

The expansion of economically important government weather forecasting services during the 1930s and their increasing need for data motivated many nations to begin regular radiosonde observation programs.

In 1985, as part of the Soviet Union's Vega program, the two Venus probes, Vega 1 and Vega 2, each dropped a radiosonde into the atmosphere of Venus. The sondes were tracked for two days.

Although modern remote sensing by satellites, aircraft and ground sensors is an increasing source of atmospheric data, none of these systems can match the vertical resolution (30 m (98 ft) or less) and altitude coverage (30 km (19 mi)) of radiosonde observations, so they remain essential to modern meteorology.

Although hundreds of radiosondes are launched worldwide each day year-round, fatalities attributed to radiosondes are rare. The first known example was the electrocution of a lineman in the United States who was attempting to free a radiosonde from high-tension power lines in 1943. In 1970 an Antonov 24 operating Aeroflot Flight 1661 suffered a loss of control after striking a radiosonde in flight resulting in the death of all 45 people on board.

Operation

A rubber or latex balloon filled with either helium or hydrogen lifts the device up through the atmosphere. The maximum altitude to which the balloon ascends is determined by the diameter and thickness of the balloon. Balloon sizes can range from 100 to 3,000 g (3.5 to 105.8 oz). As the balloon ascends through the atmosphere, the pressure decreases, causing the balloon to expand. Eventually, the balloon will expand to the extent that its skin will break, terminating the ascent. An 800 g (28 oz) balloon will burst at about 21 km (13 mi). After bursting, a small parachute on the radiosonde's support line carries it to Earth. A typical radiosonde flight lasts 60 to 90 minutes. One radiosonde from Clark Air Base, Philippines, reached an altitude of 155,092 ft (47,272 m).

The modern radiosonde communicates via radio with a computer that stores all the variables in real time. The first radiosondes were observed from the ground with a theodolite, and gave only a wind estimation by the position. With the advent of radar by the Signal Corps it was possible to track a radar target carried by the balloons with the SCR-658 radar. Modern radiosondes can use a variety of mechanisms for determining wind speed and direction, such as a radio direction finder or GPS. The weight of a radiosonde is typically 250 g (8.8 oz).

Sometimes radiosondes are deployed by being dropped from an aircraft instead of being carried aloft by a balloon. Radiosondes deployed in this way are called dropsondes.

Routine Radiosonde Launches

Worldwide there are about 1,300 radiosonde launch sites. Most countries share data with the rest of the world through international agreements. Nearly all routine radiosonde launches occur 45 minutes before the official observation time of 0000 UTC and 1200 UTC, so as to provide an instantaneous snapshot of the atmosphere. This is especially important for numerical modeling. In the United States the National Weather Service is tasked with providing timely upper-air observations for use in weather forecasting, severe weather watches and warnings, and atmospheric research. The National Weather Service launches radiosondes from 92 stations in North America and the Pacific Islands twice daily. It also supports the operation of 10 radiosonde sites in the Caribbean.

Uses of Upper Air Observations

Raw upper air data is routinely processed by supercomputers running numerical models. Forecasters often view the data in a graphical format, plotted on thermodynamic diagrams such as Skew-T log-P diagrams, Tephigrams, and or Stüve diagrams, all useful for the interpretation of the atmosphere's vertical thermodynamics profile of temperature and moisture as well as kinematics of vertical wind profile.

Radiosonde data is a crucially important component of numerical weather prediction. Because a sonde may drift several hundred kilometers during the 90- to 120-minute flight, there may be concern that this could introduce problems into the model initialization. However, this appears not to be so except perhaps locally in jet stream regions in the stratosphere.

SODAR

Wind measurement with a Phased Array SODAR

SODAR (SOnic Detection And Ranging), also written as sodar, is a meteorological instrument used as a wind profiler to measure the scattering of sound waves by atmospheric turbulence. SODAR systems are used to measure wind speed at various heights above the ground, and the thermodynamic structure of the lower layer of the atmosphere.

Sodar systems are in fact nothing more than sonar systems used in the air rather than in water; more specifically, since they operate using the Doppler effect with a multi-beam configuration to determine wind speed, they are the exact in-air equivalent to a subclass of sonar systems known as acoustic Doppler current profilers. Other names used for sodar systems include sounder, echo-sounder and acoustic radar.

Doppler Sodar

Commercial sodars operated for the purpose of collecting upper-air wind measurements consist of antennas that transmit and receive acoustic signals. A mono-static system uses the same antenna for transmitting and receiving, while a bi-static system uses separate antennas. The difference between the two antenna systems determines whether atmospheric scattering is by temperature fluctuations (in mono-static systems), or by both temperature and wind velocity fluctuations (in bi-static systems).

Mono-static antenna systems can be divided into two categories: those using multiple axis, individual antennas and those using a single phased array antenna. The multiple-axis systems generally use three individual antennas aimed in specific directions to steer the acoustic beam. Using three independent (i.e. non-colinear) axes is enough to retrieve the three components of the wind speed, although using more axes would add redundancy and increase robustness to noise when estimating the wind speed, using a least-squares approach. One antenna is generally aimed vertically, and the other two are tilted slightly from the vertical at an orthogonal angle. Each of the individual antennas may use a single transducer focused into a parabolic reflector to form a parabolic loudspeaker, or an array of speaker drivers and horns (transducers) all transmitting in-phase to form a single beam. Both the tilt angle from the vertical and the azimuth angle of each antenna are fixed when the system is set up.

Phased-array antenna systems use a single array of speaker drivers and horns (transducers), and the beams are electronically steered by phasing the transducers appropriately. To set up a phased-array antenna, the pointing direction of the array is either level, or oriented as specified by the manufacturer.

AQ500 SoDAR used in wind energy development and wind condition monitoring

The horizontal components of the wind velocity are calculated from the radially measured Doppler shifts and the specified tilt angle from the vertical. The tilt angle, or zenith angle, is generally 15 to 30 degrees, and the horizontal beams are typically oriented at right angles to one another. Since the Doppler shift of the radial components along the tilted beams includes the influence of both the horizontal and vertical components of the wind, a correction for the vertical velocity is needed in systems with zenith angles less than 20 degrees. In addition, if the system is located in a region where vertical velocities may be greater than about 0.2 m/s, corrections for the vertical velocity are needed, regardless of the beam's zenith angle.

The vertical range of sodars is approximately 0.2 to 2 kilometers (km) and is a function of frequency, power output, atmospheric stability, turbulence, and, most importantly, the noise environment in which a sodar is operated. Operating frequencies range from less than 1000 Hz to over 4000 Hz, with power levels up to several hundred watts. Due to the attenuation characteristics of the atmosphere, high power, lower frequency sodars will generally produce greater height coverage. Some sodars can be operated in different modes to better match vertical resolution and range to the application. This is accomplished through a relaxation between pulse length and maximum altitude.

Sodar Applications

Wind monitoring using the Fulcrum3D Sodar

Traditionally used in atmospheric research, sodars are now being applied as an alternative to traditional wind monitoring for the development of wind power projects. Sodars used for wind power applications are typically focused on a measurement range from 50m to 200m above ground level, corresponding to the size of modern wind turbines. Some sodar products, such as the REMTECH PA-XS Sodar and the AQ510 Sodar, have been specifically developed for this market.

Compact-beam sodars are more accurate in complex terrain where the wind vector can change across the measurement area of the sodar. By providing a more compact beam angle, these sodars reduce the effect of any change in the wind vector. This provides a more accurate estimate of wind flow and therefore energy production of a wind turbine. Compact beam sodars also reduce the effect of fixed echos and allow a more compact unit design.

Multiple-axis sodars provide the capability for simultaneous firing of all three sound beams, unlike single-axis sodars which must fire each sound beam sequentially. Simultaneous firing can provide three times the number of sample points in any given period, resulting in a higher signal to noise ratio (SNR), higher data availability and greater accuracy.

Sodars designed for the wind energy industry also differ in important aspects such as the traceability of data as a number of manufacturers do not return full signal and noise spectrum data from the sodar unit, but rather, only return processed wind speed data. This means the raw data cannot be re-analysed or reprocessed.

Analogy and Differences between Sodar and ADCP

The underlying physical principles behind the two devices are exactly the same. Both devices use sound waves to determine remote properties of the environment. Both devices use the Doppler effect to measure radial speeds on at least three non-colinear beams, which after simple computations yield the three vector components of the speed of the transmitting medium (air or water) at different altitudes. Both sodars and ADCPs can use either separate transducers for each beam or use phased arrays. Finally, both devices may use piezoelectric transducers to produce and receive the sound.

However, the operating frequencies between sodars and ADCPs are typically different. Commercial ADCPs as manufactured *e.g.* by Teledyne RDI (the *de facto* leader of this market) typically use carrier frequencies that are in the hundred of kilohertz range (300 kHz, 600 kHz, 1200 kHz) while sodars transmit only in the low kilohertz range. Transmitting at a higher frequency is possible for ADCPs due to the better sound transmitting qualities of water, and this also benefits to the compacity of the device (a diameter of typically 25 cm/10" or less for ADCPs). Also, the acoustic impedance of the transducers is not the same, because they do not operate in the same medium: air for sodars, water for ADCPs; said differently, an ADCP would not work in the air, and a sodar would not work underwater. Finally, it is more common for ADCPs to use four beams, even when they are not using a phased array. This has the benefit of adding some form of redundancy, thus making the estimate of the water currents more robust to noise. This is feasible for sodars too, but for the cost of adding a fourth transducer. The operating range of typical ADCPs is less than two hundred meters (this lowers as frequency increases, as in air).

BAROGRAPH

A traditional barograph, without its protective case

A barograph is a barometer that records the barometric pressure over time in graphical form. This instrument is also used to make a continuous recording of atmospheric pressure. The pressure-sensitive element, a partially evacuated metal cylinder, is linked to a pen arm in such a way that the vertical displacement of the pen is proportional to the changes in the atmospheric pressure.

Development

Alexander Cumming, a watchmaker and mechanic, has a claim to having made the first effective recording barograph in the 1760s using an aneroid cell. Cumming created a series of barometrical clocks, including one for King George III. However, this type of design fell out of favour. Since the amount of movement that can be generated by a single aneroid is minuscule, up to seven aneroids (so called Vidie-cans) are often stacked "in series" to amplify their motion. This type of barograph was invented in 1844 by the Frenchman Lucien Vidi (1805–1866).

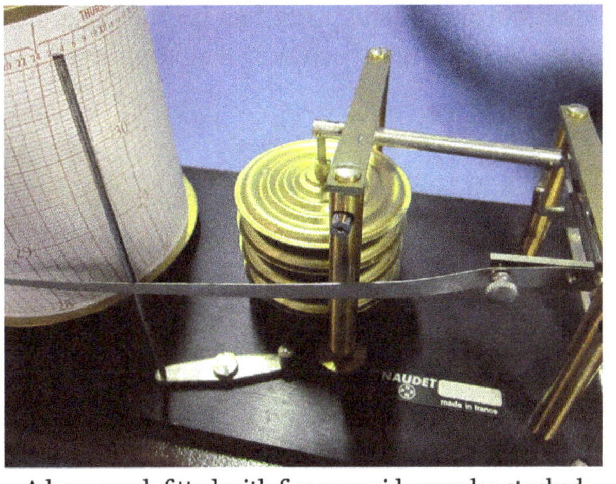

A barograph fitted with five aneroid capsules stacked in series, to amplify the amount of movement

In such barographs one or more aneroid cells act through a gear or lever train to drive a recording arm that has at its extreme end either a scribe or a pen. A scribe records on smoked foil while a pen records on paper using ink, held in a nib. The recording material is mounted on a cylindrical drum which is rotated slowly by clockwork. Commonly, the drum makes one revolution per day, per week, or per month and the rotation rate can often be selected by the user.

Various other types of barograph have also been invented. Karl Kreil described a machine in 1843 based on a syphon barometer, where a pencil marked a chart at uniform intervals. Francis Ronalds, the Honorary Director of the Kew Observatory, created the first successful barograph utilising photography in 1845. The changing height of the mercury in the barometer was recorded on a continuously moving photosensitive surface. By 1847, a sophisticated temperature-compensation mechanism was also employed. Ronalds' barograph was utilised by the UK Meteorological Office for many years to assist in weather forecasting and the machines were supplied to numerous observatories around the world.

Modern Use

Today, traditional recording barographs for meteorological use have commonly been superseded (though not all) by electronic weather instruments that use computer methods to record the barometric pressure. These are not only less expensive than earlier barographs but they may also offer both greater recording length and the ability to perform further data analysis on the captured data including automated use of the data to forecast the weather. Older mechanical barographs are highly prized by collectors as they make good display items, often being made of high quality woods and brass.

The most common weather Barograph found in homes and public buildings these days are the 8-day type. Some important manufacturers of Barographs are Negretti and Zambra, Short and Mason, and Richard Ferris among others. The late Victorian to early 20th century is generally considered to be the heyday of Barograph manufacture, many important refinements were made at this time, including improved temperature compensation and modification of the pen arm, to allow less weight to be applied to the paper, allowing better registration of small pressure changes (i.e. less friction on the nib). Marine barographs (used on ships) often include damping, this evens out the motion of the ship so that a more stable reading can be obtained, this can be either oil damping of the mechanism or simple coiled spring feet on the base. But, newer solid state, digital barographs eliminate this issue altogether, since they use no moving parts.

Use in Aviation

A Sailplane barograph in its case

As atmospheric pressure responds in a predictable manner to changes in altitude, barographs may be used to record elevation changes during an aircraft flight. Barographs were required by the FAI to record certain tasks and record attempts associated with sailplanes. A continuously varying trace indicated that the sailplane had not landed during a task, while measurements from a calibrated trace could be used to establish the completion of altitude tasks or the setting of records. Examples of FAI approved sailplane barographs included the Replogle mechanical drum barograph and the EW electronic barograph (which may be used in conjunction with GPS). Mechanical barographs are not commonly used for flight documentation now, having been displaced by GNSS Flight Recorders.

Three-day Barograph

Three-day barograph of the type used by the Meteorological Service of Canada

On the top right of the picture of the three-day barograph can be seen a silver knurled knob. This is to adjust the barograph so that it correctly reflects the station pressure. Barely visible below the knob is a small silver plunger. This is pressed every three hours to leave a time mark on the paper.

The line between two of these marks is called the 'characteristic of barometric tendency' and is used by weather forecasters. The observer would first note if the pressure was lower or higher than three hours prior. Next, a code number would be chosen that best represents the three-hour trace. There are nine possible choices (0 to 8) and no single code has preference over another. In the case of the graph on the barograph, one of two codes could be picked. An 8 (steady then decreasing) or 6 (decreasing then steady). The observer should pick the 6 because it represents the last part of the trace and is thus most representative of the pressure change.

In the bottom centre is the aneroid (large circular silver object). As the pressure increases, the aneroid is pushed down causing the arm to move up and leave a trace on the paper. As the pressure decreases, the spring lifts the aneroid and the arm moves down.

After three days the drum to which the graph is attached is removed. At this point the clockwork motor is wound and if necessary corrections can be made to increase or decrease the speed and new chart is attached.

WIND PROFILER

A wind profiler is a type of weather observing equipment that uses radar or sound waves (SODAR) to detect the wind speed and direction at various elevations above the ground. Readings are made at each kilometer above sea level, up to the extent of the troposphere (i.e., between 8 and 17 km above mean sea level). Above this level there is inadequate water vapor present to produce a radar "bounce." The data synthesized from wind direction and speed is very useful to meteorological forecasting and timely reporting for flight planning.

WindCollector2 with a Scintec SODAR SFAS

Principle

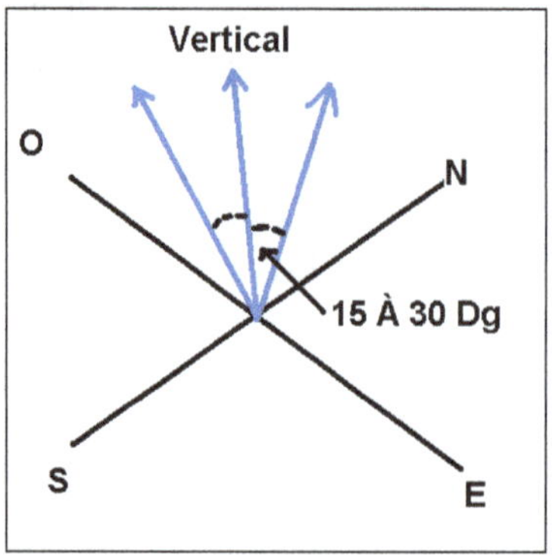

Orientation of the beams in the case of a three tilted wind profiler

In a typical implementation, the radar or sodar can sample along each of five beams: one is aimed vertically to measure vertical velocity, and four are tilted off vertical and oriented orthogonal to one another to measure the horizontal components of the air's motion. A profiler's ability to measure winds is based on the assumption that the turbulent eddies that induce scattering are carried along by the mean wind. The energy scattered by these eddies and received by the profiler is orders of magnitude smaller than the energy transmitted. However, if sufficient samples can be obtained, then the amplitude of the energy scattered by these eddies can be clearly identified above the background noise level, then the mean wind speed and direction within the volume being sampled can be determined. The radial components measured by the tilted beams are the vector sum of the horizontal motion of the air toward or away from the radar and any vertical motion present in the beam. Using appropriate trigonometry, the three-dimensional meteorological velocity components (u,v,w) and wind speed and wind direction are calculated from the radial velocities with corrections for vertical motions.

Radar Wind Profiler

A radar wind profiler

Pulse-Doppler radar wind profilers operate using electromagnetic (EM) signals to remotely sense winds aloft. The radar transmits an electromagnetic pulse along each of the antenna's pointing directions. A UHF profiler includes subsystems to control the radar's transmitter, receiver, signal processing, and Radio Acoustic Sounding System (RASS), if provided, as well as data telemetry and remote control.

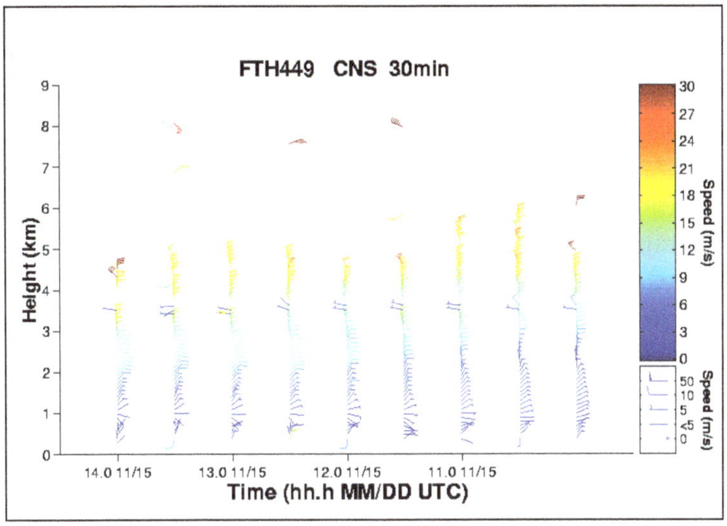

Horizontal plotted wind from a profiler

A UHF profiler includes subsystems to control the radar's transmitter, receiver, signal processing, and Radio Acoustic Sounding System (RASS), if provided, as well as data telemetry and remote control.

The duration of the transmission determines the length of the pulse emitted by the antenna, which in turn corresponds to the volume of air illuminated (in electrical terms) by the radar beam. Small amounts of the transmitted energy are scattered back (referred to as backscattering) toward and received by the radar. Delays of fixed intervals are built into the data processing system so that the radar receives scattered energy from discrete altitudes, referred to as range gates. The Doppler frequency shift of the backscattered energy is determined, and then used to calculate the velocity of the air toward or away from the radar along each beam as a function of altitude. The source of the backscattered energy (radar "targets") is small-scale turbulent fluctuations that induce irregularities in the radio refractive index of the atmosphere. The radar is most sensitive to scattering by turbulent eddies whose spatial scale is ½ the wavelength of the radar, or approximately 16 centimeters (cm) for a UHF profiler.

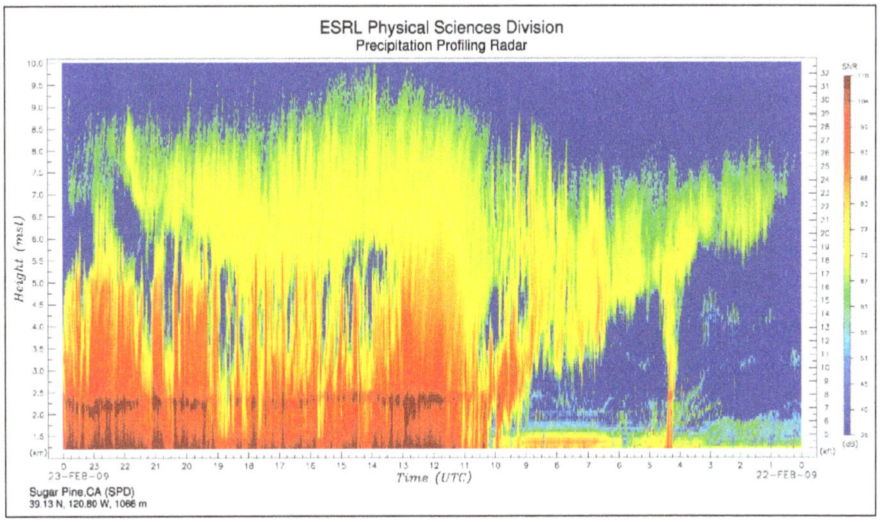

Reflectivity data obtained as a byproduct on a typical radar wind profiler

A boundary-layer radar wind profiler can be configured to compute averaged wind profiles for periods ranging from a few minutes to an hour. Boundary-layer radar wind profilers are often configured to sample in more than one mode. For example, in a "low mode," the pulse of energy transmitted by the profiler may be 60 m in length. The pulse length determines the depth of the column of air being sampled and thus the vertical resolution of the data. In a "high mode," the pulse length is increased, usually to 100 m or greater. The longer pulse length means that more energy is being transmitted for each sample, which improves the signal-to-noise ratio (SNR) of the data. Using a longer pulse length increases the depth of the sample volume and thus decreases the vertical resolution in the data. The greater energy output of the high mode increases the maximum altitude to which the radar wind profiler can sample, but at the expense of coarser vertical resolution and an increase in the altitude at which the first winds are measured. When radar wind profilers are operated in multiple modes, the data are often combined into a single overlapping data set to simplify postprocessing and data validation procedures.

Radar wind profilers may also have additional uses, for example in a biological context to complement large-scale bird monitoring schemes.

Sodar Wind Profiler

Alternatively, a wind profiler may use sound waves to measure wind speed at various heights above the ground, and the thermodynamic structure of the lower layer of the atmosphere. These sodars can be divided in mono-static system using the same antenna for transmitting and receiving, and bi-static system using separate antennas. The difference between the two antenna systems determines whether atmospheric scattering is by temperature fluctuations (in mono-static systems), or by both temperature and wind velocity fluctuations (in bi-static systems).

The TRITON transportable SODAR system used to measure wind profiles from Second Wind.

Mono-static antenna systems can be divided further into two categories: those using multiple axis, individual antennas and those using a single phased array antenna. The multiple-axis systems generally use three individual antennas aimed in specific directions to steer the acoustic beam. One antenna is generally aimed vertically, and the other two are tilted slightly from the vertical at an orthogonal angle. Each of the individual antennas may use a single transducer focused into a parabolic reflector to form a parabolic loudspeaker, or an array of speaker drivers and horns (transducers) all transmitting in-phase to form a single beam. Both the tilt angle from the vertical and the azimuth angle of each antenna are fixed when the system is set up.

The vertical range of sodars is approximately 0.2 to 2 kilometers (km) and is a function of frequency, power output, atmospheric stability, turbulence, and, most importantly, the noise environment in which a sodar is operated. Operating frequencies range from less than 1000 Hz to over 4000 Hz, with power levels up to several hundred watts. Due to the attenuation characteristics of the atmosphere, high power, lower frequency sodars will generally produce greater height coverage. Some sodars can be operated in different modes to better match vertical resolution and range to the application. This is accomplished through a relaxation between pulse length and maximum altitude.

CEILING PROJECTOR

The ceiling projector or cloud searchlight is used to measure the height of the base of clouds (called the ceiling) above the ground. It is used in conjunction with an alidade, usually positioned 1000 ft (304.8 m) away and wherever possible set at the same level.

The projector is normally set at 90°, although 71° 31' may be used, in relation to the terrain. The projector consists of a 430 W incandescent bulb set in a weatherproof housing. Inside the housing are two mirrors; the first, above the bulb, reflects the light downwards to the second mirror, that then reflects the light upwards to the cloud. Both mirrors are focused to produce a high intensity beam of light that renders a visible spot on the base of the cloud.

Ceiling projector exterior

The alidade is mounted on a post at a height of 5 ft (1.524 m) from the ground. It consists of an arm with a pointer and open sight at one end and a rubber eyepiece at the other. The arm is mounted onto a curved scale that is marked both in meters and the coded cloud height (feet). The observer looks through the eyepiece and sets the sight onto the spot projected on the cloud and reads the height from the attached scale.

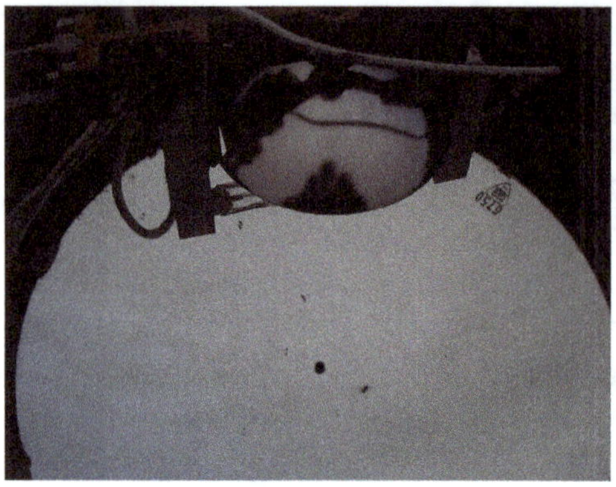

Interior of a ceiling projector

Alidade for use with a ceiling projector

When the cloud is thin the beam of light may penetrate into the cloud. The observer should read the scale where the light first enters the cloud and not at the top. However, a remark may be made as to how far into the cloud the light was able to penetrate as this may be useful. In the case of fog or blizzard conditions the observer should read the scale where the beam disappears.

DARK ADAPTOR GOGGLES

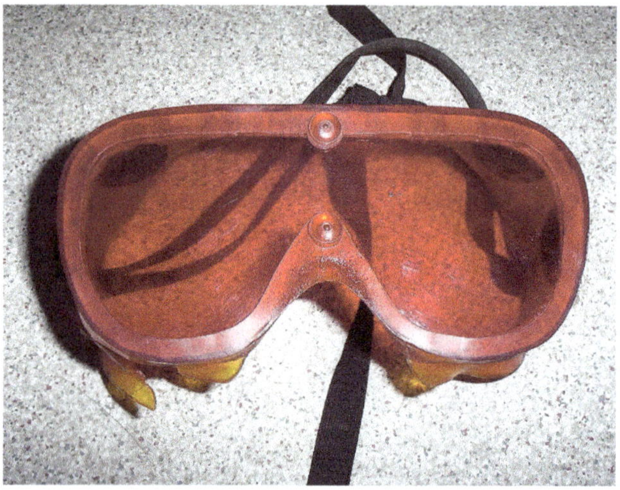

Dark adaptor goggles

Dark adaptor goggles, also called red adaptation goggles, are used in the field of meteorology and astronomy for adapting the eyes to the dark prior to an observation at night. They also aid with the identification of clouds during bright sunshine or glare from snow. The goggles are made with red-tinted plastic lenses. Such goggles or glasses are often used by pilots and weather observers to preserve their natural night vision.

To help with the identification of clouds during the day the goggles should be put on just before the observer goes outside. This will ensure the eyes are not affected by the glare from the sun or snow. For use at night the observer should put the goggles on about ten minutes prior to going outside. Once the observer is outside and away from major sources of light the goggles may be removed and the observer's vision should be adapted to the dark. Although the goggles are designed to be worn indoors, some people find walking around inside, especially up and down stairs, to be difficult. At no time should the goggles be used to help assess the visibility.

Dark adaptor goggles were invented by Wilhelm Trendelenburg in 1916. The concept is based on the work by Antoine Béclère on dark adaptation of the eye, where it is noted that fluoroscopy relies on the use of the retinal rods of the eye. Since retinal rods are insensitive to long wavelengths of light, such as red light, while retinal cones are not, the goggles enabled the physicians to adapt their eyes in preparation for the fluoroscopic procedure while still being able to perform other work. Prior to the invention of these goggles, physicians were required to sit in the darkened room where the procedure would be performed for extended periods to adapt their eyes to the low lighting conditions.

WINDSOCK

Windsock

A windsock is a conical textile tube that resembles a giant sock. Windsocks can be used as a basic guide to wind direction and speed, or as decoration.

Uses

Windsocks are used to tell wind speed and the direction of the wind speed itself. Windsocks typically are used at airports to indicate the direction and strength of the wind to pilots and at chemical plants where there is risk of gaseous leakage. They are sometimes located alongside highways at windy locations.

At many airports, windsocks are lit at night, either by floodlights on top surrounding it or with one mounted on the pole shining inside it.

Function

Wind direction is the opposite of the direction in which the windsock is pointing (note that wind directions are conventionally specified as being the compass point from which the wind originates;

so a windsock pointing due north indicates a southerly wind). Wind speed is indicated by the windsock's angle relative to the mounting pole; in low winds, the windsock droops; in high winds it flies horizontally.

Alternating stripes of high visibility orange and white were initially used to help to estimate the speed of wind. Each stripe adds up 3 knots to the estimated wind speed. However, some circle frames mountings cause windsocks to be held open at one end, indicating a velocity of 3 knots, even though anemometers would show no wind speed. A fully extended windsock suggests a wind speed of 15 knots (28 km/h; 17 mph) or greater.

SOLARIMETER

A solarimeter is a pyranometer, a type of measuring device used to measure combined direct and diffuse solar radiation. An integrating solarimeter measures energy developed from solar radiation based on the absorption of heat by a black body. The principle this instrument was designed on was first developed by the Italian priest, Father Angelo Bellani. He invented the actinometric method which is based on physical and chemical techniques.

NEPHOSCOPE

A nephoscope was an instrument for measuring the altitude, direction, and velocity of clouds in the 19th century. A ray of light of known velocity is emitted from a nephoscope, which strikes the base of a targeted cloud. The travel time of the return signal is used to estimate the distance to the cloud.

A nephoscope was an instrument used to measuring the altitude, direction, and velocity of clouds in the 19th century. This is different from a "nephometer" which is an instrument used in measuring the amount of cloudiness.

A ray of light of known velocity is emitted from a nephoscope, which strikes the base of a targeted cloud. The travel time of the return signal is used to estimate the distance to the cloud.

The distance to the cloud is calculated *approximately* using the equation:

Distance = (speed of light × duration of delay due to reflection)/2.

Mirror Nephoscope

Developed by Carl Gottfrid Fineman this instrument consists of a magnetic compass, the case of which is covered with a black mirror, around which is movable a circular metal frame. A little window in this mirror enables the observer to see the tip of the compass needle underneath. On the surface of the mirror are engraved three concentric circles and four diameters; one of the latter passes through the middle of the little window. The mirror constitutes a compass card, its radii corresponding to the cardinal points. On the movable frame surrounding. The mirror is fixed a vertical pointer graduated in millimeters, which can be moved· up and down by means of a rack and pinion. The whole apparatus is mounted on a tripod stand provided with leveling screws. To make an observation, the mirror is adjusted to the horizontal with the leveling-screws, and is oriented to the meridian by moving the whole apparatus until the compass needle is seen, through the window, to lie in the north-south line of the mirror (making, however, allowance for the magnetic declination). The observer stands in such a position as to bring the image of any chosen part of a cloud at the center of the mirror, lind the vertical pointer is also adjusted by screwing it up or down and by rotating it around the mirror until its tip is reflected in the center' of the mirror. As the image of the cloud moves toward the circumference of the mirror the observer moves his head so as to keep the tip of the pointer and the cloud image in coincidence. The radius along which the image moves gives the direction of the cloud's movement, and the time required to pass from one circle to the next its relative speed, which may be reduced to certain arbitrary units. This instrument is, however, not very easy to use, and gives only moderately accurate measurements.

Mirror Nephoscope

Comb Nephoscope

Besson comb nephoscope

Developed by Louis Besson director of the Observatory of Montsouris in 1912 this apparatus consists of a horizontal bar fitted with several equidistant spikes and mounted on the upper end of a vertical pole which can be rotated on its axis. When an observation is to be made the observer places himself in such a position that the central spike is projected on any chosen part of a cloud. Then, without altering his position, he causes the "comb" to turn by means of two cords in such a manner that the cloud is seen to follow along the line of spikes. A graduated circle, turning with the vertical pole, gives the direction of the cloud's motion ; it is read with the aid of a fixed pointer. Moreover, when the apparatus is once oriented, the observer can determine the relative speed of the cloud by noting the time the latter requires to pass from one spike to the next. If the instrument stands on level ground, so that the observer's eye is always at the same height, and if the interval between two successive spikes is equal to one tenth of their altitude above the eye-level of the observer, one need only multiply by 10 the time required for the cloud to pass over one interval to determine the time the cloud travels a horizontal distance equal to its altitude.

M. Besson revived an old method of Bravais for measuring the actual height of clouds. The apparatus in this case consists of a plate of glass having parallel faces, mounted on a graduated vertical circle which indicates its angle of inclination. A sheet of water, situated at a lower level, serves as a mirror to reflect the cloud. The water is contained in a reservoir of blackened cement surrounded by shrubbery, and is only a small fraction of an inch in depth, so that the wind may not disturb its level surface. The observer, having mounted' the glass plate on the horizontal axis of a theodolite set on a window-sill some 30 or 40 feet above the ground, places his eye close to it and adjusts its inclination so that the images of a cloud reflected in the plate and in the sheet of water coincide. Then from a curve traced once for all on a sheet of plotting paper he reads off the altitude of the cloud corresponding to the observed angle on the glass plate. The curve is plotted from simple trigonometrical calculations. At the Observatory of Montsouris the degree of cloudiness, i. e., the amount of the whole sky covered with clouds at a given moment, is determined by means of the nephometer, also devised by M. Besson. This consists of a convex glass mirror, a segment of a sphere, about twelve inches in diameter, in which is seen the reflection of the celestial vault divided into ten sections of equal area by means of lines engraved on the glass. As shown, the meteorologist observes through an eyepiece fixed in an invariable position with respect to the mirror, which latter turns freely on a vertical axis. The observer, whose own image partly obstructs sections 8, 9. and 10, notes the degree of cloudiness in the sections numbered 1 to 7. The cloudiness of each section is estimated on a scale of 0 to 10 ; zero meaning no clouds and 10 entirely overcast. He now rotates the mirror and eyepiece 180 degrees and observes the cloudiness in sections 7, 5, and 2, which represent the regions of the sky that at the first observation corresponded to sections 8, 9, and 10.

LIGHTNING DETECTION

A lightning detector is a device that detects lightning produced by thunderstorms. There are three primary types of detectors: *ground-based* systems using multiple antennas, *mobile systems* using a direction and a sense antenna in the same location (often aboard an aircraft), and *space-based systems*.

The first such device was invented in 1894 by Alexander Stepanovich Popov. It also was the first radio receiver in the world.

Lightning detector at the Kennedy Space Center in Florida

Ground-based and mobile detectors calculate the direction and severity of lightning from the current location using radio direction-finding techniques along with an analysis of the characteristic frequencies emitted by lightning. Ground-based systems use triangulation from multiple locations to determine distance, while mobile systems estimate distance using signal frequency and attenuation. Space-based detectors on satellites can be used to locate lightning range, bearing and intensity by direct observation.

Ground-based lightning detector networks are used by meteorological services like the National Weather Service in the United States, the Meteorological Service of Canada, the European Cooperation for Lightning Detection (EUCLID), the Institute for Ubiquitous Meteorology (Ubimet) and by other organizations like electrical utilities and forest fire prevention services.

Limitations

Each system used for lightning detection has its own limitations. These include:

- A ground-based lightning network must be able to detect a flash with at least three antennas to locate it with an acceptable margin of error. This often leads to the rejection of cloud-to-cloud lightning, as one antenna might detect the position of the flash on the starting cloud and the other antenna the receiving one. As a result, ground-based networks have a tendency to underestimate the number of flashes, especially at the beginning of storms where cloud-to-cloud lightning is prevalent.
- Since they use attenuation rather than triangulation, mobile detectors sometimes mistakenly indicate a weak lightning flash nearby as a strong one further away, or vice versa.
- Space-based lightning networks suffer from neither of these limitations, but the information provided by them is often several minutes old by the time it is widely available, making it of limited use for real-time applications such as air navigation.

Aviation use

Large airliners are more likely to use weather radar than lightning detectors, since weather radar can detect smaller storms that also cause turbulence; however, modern avionics systems often include lightning detection as well, for additional safety.

For smaller aircraft, especially in general aviation, there are two main brands of lightning detectors (often referred to as sferics, short for radio atmospherics): *Stormscope*, produced originally by Ryan (later B.F. Goodrich) and currently by L-3 Communications, and the *Strikefinder*, produced by Insight. *Strikefinder* can detect and properly display IC (intracloud) and CG (cloud to ground) strikes as well as being able to differentiate between real strikes and signal bounces reflected off the Ionosphere. Lightning detectors are inexpensive and lightweight, making them attractive to owners of light aircraft (particularly of single-engine aircraft, where the aircraft nose is not available for installation of a radome).

Professional-quality Portable Lightning Detectors

Lightning strike counter in a Museum Patio

Inexpensive portable lightning detectors as well as other single sensor lightning mappers, such as used on aircraft, have limitations including detection of false signals and poor sensitivity, particularly for intracloud (IC) lightning. Professional-quality portable lightning detectors improve performance in these areas by several techniques which facilitate each other, thus magnifying their effects:

- False signal elimination: A lightning discharge generates both a radio frequency (RF) electromagnetic signal – commonly experienced as "static" on an AM radio – and very short duration light pulses, comprising the visible "flash". A lightning detector that works by sensing just one of these signals may misinterpret signals coming from sources other than lightning, giving a false alarm. Specifically, RF-based detectors may misinterpret RF noise, also known as RF Interference or RFI. Such signals are generated by many common environmental sources, such as auto ignitions, fluorescent lights, TV sets, light switches, electric motors, and high voltage wires. Likewise, light-flash-based detectors may misinterpret flickering light generated in the environment, such as reflections from windows, sunlight through tree leaves, passing cars, TV sets, and fluorescent lights.

However, since RF signals and light pulses rarely occur simultaneously except when produced by lightning, RF sensors and light pulse sensors can usefully be connected in a "coincidence circuit" which requires both kinds of signals simultaneously in order to produce an output. If such a system is pointed toward a cloud and lightning occurs in that cloud, both signals will be received; the coincidence circuit will produce an output; and the user can be sure the cause was lightning. When a lightning discharge occurs within a cloud at night, the entire cloud appears to illuminate. In daylight these intracloud flashes are rarely visible to the human eye; nevertheless, optical sensors can detect them. Looking through the window of the space shuttle in early missions, astronauts used optical sensors to detect lightning in bright sunlit clouds far below. This application led to development of the dual signal portable lightning detector which utilizes light flashes as well as the "sferics" signals detected by previous devices.

- Improved Sensitivity: In the past, lightning detectors, both inexpensive portable ones for use on the ground and expensive aircraft systems, detected low frequency radiation because at low frequencies the signals generated by cloud-to-ground (CG) lightning are stronger (have higher amplitude) and thus are easier to detect. However, RF noise is also stronger at low frequencies. To minimize RF noise reception, low-frequency sensors are operated at low sensitivity (signal reception threshold) and thus do not detect less intense lightning signals. This reduces the ability to detect lightning at longer distances since signal intensity decreases with the square of distance. It also reduces detection of intracloud (IC) flashes which generally are weaker than CG flashes.
- Enhanced Intracloud Lightning Detection: The addition of an optical sensor and coincidence circuit not only eliminates false alarms caused by RF noise; it also allows the RF sensor to be operated at higher sensitivity and to sense higher frequencies characteristic of IC lightning and enable the weaker high frequency components of IC signals and more distant flashes to be detected.

The improvements described above significantly extend the detector's utility in many areas:

- Early warning: Detection of IC flashes is important because they typically occur from 5 to 30 minutes before CG flashes and so can provide earlier warning of developing thunderstorms, greatly enhancing the effectiveness of the detector in personal-safety and storm-spotting applications compared to a CG-only detector. Increased sensitivity also provides warning of already-developed storms which are more distant but may be moving toward the user.
- Storm location: Even in daylight, "storm chasers" can use directional optical detectors that can be pointed at an individual cloud to distinguish thunderclouds at a distance. This is particularly important for identifying the strongest thunderstorms which produce tornadoes, since such storms produce higher flash rates with more high frequency radiation than weaker non-tornadic storms.
- Microburst prediction: IC flash detection also provides a method for predicting microbursts. The updraft in convective cells starts to become electrified when it reaches altitudes sufficiently cold so that mixed phase hydrometeors (water and ice particles) can exist in the same volume. Electrification occurs due to collisions between ice particles and water drops or water coated ice particles. The lighter ice particles (snow) are charged positively and carried to the upper portion of the cloud leaving behind the negatively charged water drops in the central part of the cloud. These two charge centers create an electric field leading to lightning formation. The updraft continues until all the liquid water is converted to ice, which releases latent heat driving the updraft. When all the water is converted, the updraft collapses rapidly as does the lightning rate. Thus the increase in lightning rate to

a large value, mostly due to IC discharges, followed by a rapid dropoff in rate provides a characteristic signal of the collapse of the updraft which carries particles downward in a downburst. When the ice particles reach warmer temperatures near cloudbase they melt causing atmospheric cooling; likewise, the water drops evaporate, also causing cooling. This cooling increases air density which is the driving force for microbursts. The cool air in "gust fronts" often experienced near thunderstorms is caused by this mechanism.

- Storm identification/tracking: Some thunderstorms, identified by IC detection and observation, make no CG flashes and would not be detected with a CG sensing system. IC flashes also are many times as frequent as CG so provide a more robust signal. The relative high density (number per unit area) of IC flashes allows convective cells to be identified when mapping lightning whereas CG lightning are too few and far between to identify cells which typically are about 5 km in diameter. In the late stages of a storm the CG flash activity subsides and the storm may appear to have ended—but generally there still is IC activity going on in the residue mid-altitude and higher cirrus anvil clouds, so the potential for CG lightning still exists.
- Storm intensity quantification: Another advantage of IC detection is that the flash rate (number per minute) is proportional to the 5th power of the convective velocity of the updrafts in the thundercloud. This non-linear response means that a small change in cloud height, hardly observable on radar, would be accompanied by a large change in flash rate. For example, a hardly noticeable 10% increase in cloud height (a measure of storm severity) would have a 60% change in total flash rate, which is easily observed. "Total lightning" is both the generally invisible (in daylight) IC flashes that stay within the cloud as well as the generally visible CG flashes that can be seen extending from cloud base to ground. Because most of the total lightning is from IC flashes, this ability to quantify storm intensity occurs mostly through detection of IC discharges. Lightning detectors that sense only low frequency energy detect only IC flashes that are nearby, so they are relatively inefficient for predicting microbursts and quantifying convective intensity.
- Tornado Prediction: Severe storms that produce tornadoes are known to have very high lightning rates and most lightning from the deepest convective clouds is IC, therefore the ability to detect IC lightning provides a method for identifying clouds with high tornado potential.

Lightning Range Estimation

When an RF lightning signal is detected at a single location, one can determine its direction using a crossed-loop magnetic direction finder but it is difficult to determine its distance. Attempts have been made using the amplitude of the signal but this does not work very well because lightning signals greatly vary in their intensity. Thus, using amplitude for distance estimation, a strong flash may appear to be nearby and a weaker signal from the same flash – or from a weaker flash from the same storm cell – appears to be farther away. One can tell where lightning will strike within a mile radius by measuring ionization in the air to improve the accuracy of the prediction.

To understand this aspect of lightning detection one needs to know that a lightning 'flash' generally consists of several strokes, a typical number of strokes from a CG flash is in the range 3 to 6 but some flashes can have more than 10 strokes. The initial stroke leaves an ionized path from the cloud to ground and subsequent 'return strokes', separated by an interval of about 50 milliseconds, go up that channel. The complete discharge sequence is typically about ½ second in duration while the duration of the individual strokes varies greatly between 100 nanoseconds and a few tens of microseconds. The strokes in a CG flash can be seen at night as a non-periodic sequence of illuminations of the lightning channel. This can also be heard on sophisticated lightning detectors as individual staccato sounds for each stroke, forming a distinctive pattern.

Single sensor lightning detectors have been used on aircraft and while the lightning direction can be determined from a crossed loop sensor, the distance can not be determined reliably because the signal amplitude varies between the individual strokes described above, and these systems use amplitude to estimate distance. Because the strokes have different amplitudes, these detectors provide a line of dots on the display like spokes on a wheel extending out radially from the hub in the general direction of the lightning source. The dots are at different distances along the line because the strokes have different intensities. These characteristic lines of dots in such sensor displays are called "radial spread". These sensors operate in the very low frequency (VLF) and low frequency (LF) range (below 300 kHz) which provides the strongest lightning signals: those generated by return strokes from the ground. But unless the sensor is close to the flash they do not pick up the weaker signals from IC discharges which have a significant amount of energy in the high frequency (HF) range (up to 30 MHz).

Another issue with VLF lightning receivers is that they pick up reflections from the ionosphere so sometimes can not tell the difference in distance between lightning 100 km away and several hundred km away. At distances of several hundred km the reflected signal (termed the "sky wave") is stronger than the direct signal (termed the "ground wave").

The Earth-ionosphere waveguide traps electromagnetic VLF- and ELF waves. Electromagnetic pulses transmitted by lightning strikes propagate within that waveguide. The waveguide is dispersive, which means that their group velocity depends on frequency. The difference of the group time delay of a lighting pulse at adjacent frequencies is proportional to the distance between transmitter and receiver. Together with the direction finding method, this allows locating lightning strikes by a single station up to distances of 10000 km from their origin. Moreover, the eigenfrequencies of thc Earth-ionospheric waveguide, the Schumann resonances at about 7.5 Hz, are used to determine the global thunderstorm activity.

Because of the difficulty in obtaining distance to lightning with a single sensor, the only current reliable method for positioning lightning is through interconnected networks of spaced sensors covering an area of the Earth's surface using time-of-arrival differences between the sensors and/or crossed-bearings from different sensors. Several such national networks currently operating in the U.S. can provide the position of CG flashes but currently cannot reliably detect and position IC flashes. There are a few small area networks that have VHF time of arrival systems and can detect and position IC flashes. These are called lightning mapper *arrays*. They typically cover a circle 30–40 miles in diameter.

RADAR

Radar stands for Radio Detection and Ranging System. It is basically an electromagnetic system used to detect the location and distance of an object from the point where the Radar is placed. It works by radiating energy into space and monitoring the echo or reflected signal from the objects. It operates in the UHF and microwave range.

The Radar system generally consists of a transmitter which produces an electromagnetic signal which is radiated into space by an antenna. When this signal strikes any object, it gets reflected or reradiated in many directions. This reflected or echo signal is received by the radar antenna which delivers it to the receiver, where it is processed to determine the geographical statistics of the object. The range is determined by the calculating the time taken by the signal to travel from the Radar to the target and back. The target's location is measured in angle, from the direction of maximum amplitude echo signal, the antenna points to. To measure range and location of moving objects, Doppler Effect is used.

Radar System

Given below are 6 major parts of a Radar System:

- A Transmitter: It can be a power amplifier like a Klystron, Travelling Wave Tube or a power Oscillator like a Magnetron. The signal is first generated using a waveform generator and then amplified in the power amplifier.
- Waveguides: The waveguides are transmission lines for transmission of the Radar signals.
- Antenna: The antenna used can be a parabolic reflector, planar arrays or electronically steered phased arrays.
- Duplexer: A duplexer allows the antenna to be used as a transmitter or a receiver. It can be a gaseous device that would produce a short circuit at the input to the receiver when transmitter is working.
- Receiver: It can be super heterodyne receiver or any other receiver which consists of a processor to process the signal and detect it.
- Threshold Decision: The output of the receiver is compared with a threshold to detect the presence of any object. If the output is below any threshold, the presence of noise is assumed.

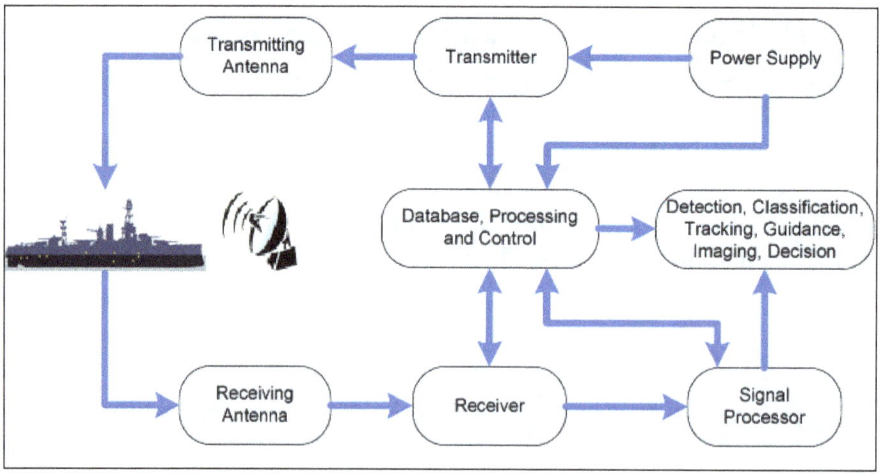

Pulsed Radar

Pulsed RADAR sends high power and high frequency pulses towards the target object. It then waits for the echo signal from the object before another pulse is send. The range and resolution of the RADAR depends on the pulse repetition frequency. It uses the Doppler shift method.

The principle of RADAR detecting moving objects using the Doppler shift works on the fact that echo signals from stationary objects are in same phase and hence get cancelled while echo signals from moving object will have some changes in phase.

Types of Pulsed Radar

Pulse Doppler RADAR: It transmits high pulse repetition frequency to avoid Doppler ambiguities. The transmitted signal and the received echo signal are mixed in a detector to get the Doppler shift and the difference signal is filtered using a Doppler filter where the unwanted noise signals are rejected.

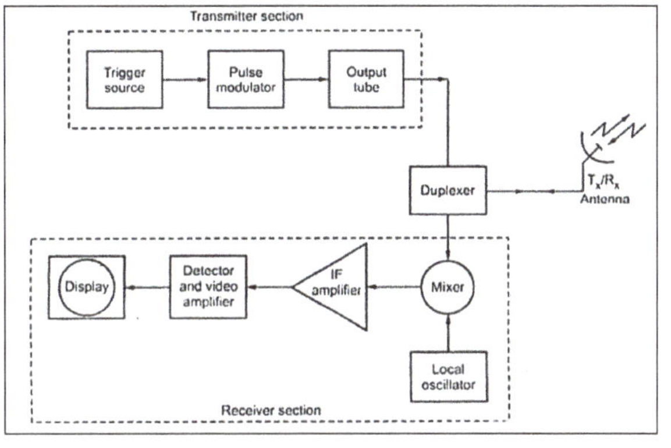

Moving Target Indicator RADAR: It transmits low pulse repetition frequency to avoid range ambiguities. In a MTI RADAR system, the received echo signals from the object are directed towards the mixer, where they are mixed with the signal from a stable local oscillator (STALO) to produce the IF signal. This IF signal is amplified and then given to the phase detector where its phase is compared with the phase of the signal from the Coherent Oscillator (COHO) and the difference signal is produced. The Coherent signal has the same phase as the transmitter signal. The coherent signal and the STALO signal are mixed and given to the power amplifier which is switched on and off using the pulse modulator.

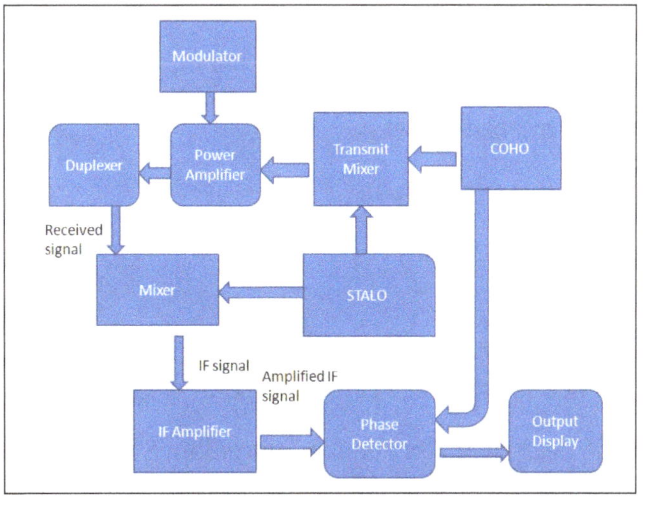

Continuous Wave Radar

The continuous wave RADAR doesn't measures the range of the target but rather the rate of change of range by measuring the Doppler shift of the return signal. In a CW RADAR electromagnetic radiation is emitted instead of pulses. It is basically used for speed measurement.

The RF signal and the IF signal are mixed in the mixer stage to generate the local oscillator frequency. The RF signal is the transmitted signal and the received signal by the RADAR antenna consists of the RF frequency plus the Doppler shift frequency. The received signal is mixed with the local oscillator frequency in the second mixture stage to generate the IF frequency signal. This signal is amplified and given to the third mixture stage where it is mixed with the IF signal to get the signal with Doppler frequency. This Doppler frequency or Doppler shift gives the rate of change of range of the target and thus the velocity of the target is measured.

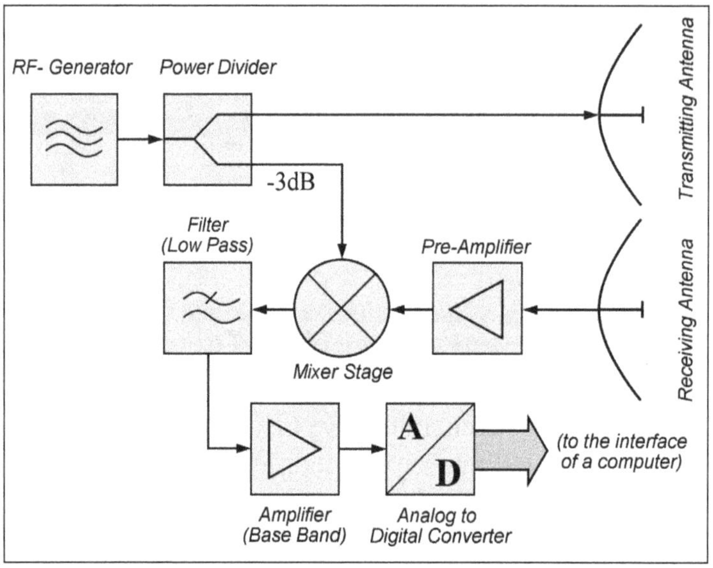

Radar Applications

Military Applications

The Radar has 3 major applications in Military:

- In air defense it is used for target detection, target recognition and weapon control (directing the weapon to the tracked targets).
- In missile system to guide the weapon.
- Identifying enemy locations in map.

Air Traffic Control

The Radar has 3 major applications in Air Traffic control:

- To control air traffic near airports. The Air Surveillance Radar is used to detect and display the aircraft's position in the airport terminals.

- To guide the aircraft to land in bad weather using Precision Approach Radar.
- To scan the airport surface for aircraft and ground vehicle positions.

Remote Sensing: Radar can be used for observing weather or observing planetary positions and monitoring sea ice to ensure smooth route for ships.

Ground Traffic Control: Radar can also be used by traffic police to determine speed of the vehicle, controlling the movement of vehicles by giving warnings about presence of other vehicles or any other obstacles behind them.

Space

Radar has 3 major applications:

- To guide the space vehicle for safe landing on moon.
- To observe the planetary systems.
- To detect and track satellites.
- To monitor the meteors.

References

- Hygrometer-kat, test-meters, measuring-instruments, English, pce-instruments.com, Retrieved 6 February, 2019
- World Meteorological Organization. "Vane anemometer". Eumetcal. Archived from the original on 8 April 2014. Retrieved 6 April 2014
- What-is-a-hygrometer, news, kestrelmeters.com, Retrieved 7 March, 2019
- Shea, William R. (2003). Designing Experiments & Games of Chance: The Unconventional Science of Blaise Pascal. Science History Publications. pp. 21–. ISBN 978-0-88135-376-1. Retrieved 10 October 2012
- How-a-hygrometer-works: lumacomfort.com, Retrieved 8 April, 2019
- Weisshaupt, N.; Arizaga, J.; Maruri, M. (2018). "The role of radar wind profilers in ornithology". Ibis. 160 (3): 516–527. doi:10.1111/ibi.12562
- Radar-basics-types-and-applications: elprocus.com, Retrieved 8 April, 2019
- Ronalds, B.F. (2016). "The Beginnings of Continuous Scientific Recording using Photography: Sir Francis Ronalds› Contribution". European Society for the History of Photography. Retrieved 2 June 2016

Applications of Meteorology

CHAPTER 6

Meteorology is applied in various sectors such as aviation, construction and energy sector. It is also used to study areas of disturbed weather in order to recognize potential tropical cyclones. The topics elaborated in this chapter will help in gaining a better perspective about these applications of meteorology.

AVIATION METEOROLOGY

Aviation meteorology is the study of weather from the unique perspective of the aviation industry. This subject began during the era of hot-air balloons and gliders. Notable early "aerologists" Léon Teisserenc de Bort of France and Robert Assmann of Germany are co-credited with the discovery of the stratosphere in 1902 via independent, multiple high-altitude balloon ascents. Across the Atlantic, aerologist Clarence LeRoy Meisinger received one of the early doctorates in the United States in meteorology in 1922, but he perished in a balloon accident during a thunderstorm just two years later. The surge in heavier-than-air powered flight during the two world wars led to the professionalization of aviation meteorology. Subjects such as high-level or "clear-air" turbulence materialized as aircraft regularly penetrated largely unexplored regions of the atmosphere. The rise of commercial aviation after World War II led to the development of private-sector aviation meteorology groups within individual airlines. Much of the development of modern aviation meteorology took place within these groups and in conjunction with government agencies, rather than in academic settings. Despite the later consolidation or elimination of many airline meteorology groups, aviation meteorology remains an unusually practical and applied specialty.

As a result, a newcomer to aviation meteorology from an academic background may be surprised at the relative lack of scholarly material compared to other subjects. Another surprise may be the distinctively empirical approach to forecast problems that still remains—although the latter has lessened in recent decades via contributions from research scientists. One enduring challenge for aviation forecasting has been that advances in prediction are often reliant on aircraft encounters with aviation hazards such as turbulence, wind shear, icing, and volcanic ash. A more desirable outcome for pilots, passengers, and airlines would be the avoidance of these hazards. As aviation forecasting has improved such encounters have decreased, which can have an ironic diminishing returns effect on forecast improvements. This limitation can be overcome to some extent by theoretical and modeling advances. The relatively new field of space weather also intersects with aviation meteorology because of the impact that an active Sun can have on aviation communication and navigation, as well as human health. Beyond the focus on forecasting for commercial aviation, environmental impacts due to contrails and climate change have been a part of aviation meteorology for several decades as well. Despite its empirical roots, aviation meteorology should not be viewed as a less worthy subject for academic study. The practical demands of aviation safety intersect with some of the most complicated phenomena and most vexing forecast problems in the entire field of meteorology, on scales that are still not resolvable by the world's best weather forecasting models. Study of aviation meteorology is rewarding: in terms both of saving lives and of reducing injuries and as a scientific pursuit in its own right.

INVEST METEOROLOGY

An invest on June 15, 2015, which later became Tropical Storm Bill. Note that it already had a cyclonic shape

An invest in meteorology (short for *investigative area*, alternatively written INVEST) is a designated area of disturbed weather that is being monitored for potential tropical cyclone development. Invests are designated by three separate United States forecast centers: the National Hurricane Center, the Central Pacific Hurricane Center, and the Joint Typhoon Warning Center.

Invests (also called *areas of interest*) are designated by three separate forecast centers located in the United States: the National Hurricane Center in Miami, Florida, overseeing the North Atlantic and North Eastern Pacific basins (east of the 140°W meridian); the Central Pacific Hurricane Center in Honolulu, Hawaii, monitoring the North Central Pacific basin (between the International Date Line and the 140°W meridian); and the Joint Typhoon Warning Center in Pearl Harbor, Hawaii (formerly located in the island of Guam), serving U.S. government interests elsewhere (like, i.e. the North Western Pacific basin west of the International Date Line). The designation of a system as an invest does not necessarily correspond to any particular likelihood of development of the system into a tropical cyclone (tropical depression, tropical storm, or hurricane/typhoon).

Designation

<table>
<tr><th colspan="4">hideINVEST Numbering (NHC/CPHC/JTWC)</th></tr>
<tr><th>Basin(s)</th><th>Warning Center</th><th>Suffix</th><th>Example</th></tr>
<tr><td>N Atlantic</td><td rowspan="2">NHC</td><td>L</td><td>Invest-90L</td></tr>
<tr><td>NE Pacific
(E of 140°W)</td><td>E</td><td>Invest-91E</td></tr>
<tr><td>NC Pacific
(E of IDL, W of 140°W)</td><td>CPHC</td><td>C</td><td>Invest-92C</td></tr>
<tr><td>NW Pacific
(W of IDL)</td><td>JTWC</td><td>W</td><td>Invest-93W</td></tr>
<tr><td>N Indian O.
(Bay of Bengal)</td><td rowspan="4">JTWC
(unoff.)</td><td>B</td><td>Invest-94B</td></tr>
<tr><td>N Indian O.
(Arabian Sea)</td><td>A</td><td>Invest-95A</td></tr>
<tr><td>SW Indian O. & Australian reg.
(W of 135°E)</td><td>S</td><td>Invest-96S</td></tr>
<tr><td>Australian reg. & S Pacific
(E of 135°E)</td><td>P</td><td>Invest-97P</td></tr>
<tr><td>S Atlantic</td><td>NRL, NHC</td><td>Q</td><td>Invest-98Q</td></tr>
</table>

Invests are numbered from 90 to 99, followed by a suffix letter "L" in the North Atlantic basin, "E" and "C" in the Eastern and Central Pacific basins (respectively), or "W" in the Western Pacific basin.

The Joint Typhoon Warning Center also issues unofficial warnings for U.S. government interests (predominantly military) in the Southern Hemisphere, designating tropical invests with the "S" suffix when they form west of 135°E (this spans the whole South Indian Ocean, including the South-Western Indian Ocean basin and the western half of the Australian-region basin), and the "P" suffix when they form east of 135°E (spans both the eastern half of the Australian region and the South Pacific basin). In addition, invests in the North Indian Ocean cyclone basin are also labelled by the JTWC, and are suffixed with "A" if they form in the Arabian Sea and with "B" if they form in the Bay of Bengal.

The Naval Research Laboratory's Marine Meteorology Division also uses the "Q" suffix to designate invests which form in the South Atlantic Ocean, even though it is not recognized as an official tropical cyclone basin by the World Meteorological Organization.

These suffix letters (with the usual exception of "L") are also used with the *tropical cyclone numbers* (*TC numbers* for short) assigned to tropical and subtropical depressions (and potential tropical cyclones) monitored by the NHC & CPHC (North Atlantic systems are usually designated by TC numbers without a suffix letter; the "L" is still used, however, by the NRL for this basin) and all tropical, subtropical, and potential tropical cyclones tracked by the JTWC.

Rotation

Numbers are rotated within the season and are re-used as necessary (the next invest after 99 would be numbered 90). In contrast, TC numbers (which start each year/season with 01 and run upwards, usually up to 30 in NHC and CPHC-monitored basins; the hard-coded limit imposed by the Automated Tropical Cyclone Forecasting System is 49) assigned to proper cyclones (tropical, subtropical, and potential tropical) are not recycled until the following year/season. If the invest system develops into a (sub)tropical cyclone, it is reclassified as the next name/number on the list (a TC number if evolving into a depression or JTWC cyclone; or a name if evolving quickly into a tropical or subtropical storm, bypassing the depression stage.

METEOROLOGY AND THE ENERGY SECTOR

Weather forecasts have long been important to the energy sector. Traditionally they have been used to anticipate demand. Hot or cold weather can lead to significant spikes in consumption caused by increased air conditioning and heating. Advance knowledge of the weather helps energy companies to predict such surges and to make sure they have adequate supplies.

As the share of solar and wind power in the energy mix grows, forecasting supply as well as demand is becoming increasingly important. The reason is that the power output from these sources of energy is highly variable. To a large extent, this variability depends on the weather.

Impact of Weather Conditions

Consider the example of wind: the electricity generated by a wind turbine may vary significantly from one hour, day, week or month to the next, depending on the direction and strength of the wind. Over longer timescales, climate variability and climate change may affect the amount of wind power available at particular sites.

Different aspects of wind information are important to different players in the energy sector:

- Policy makers need to be able to weigh the potential of wind power as part of the overall energy mix.
- Wind farm operators considering siting options need to know how much power could be produced in particular locations, and how variable that power would be.
- Grid operators need to be able to estimate how much power will be generated from wind in the hours to come so they can balance the network.
- Energy traders need to have an idea of the amount of power they will be able to trade on electricity markets on a given day.
- Wind turbine operators need to know when winds are forecast to be low to plan routine maintenance.

Similar considerations apply in the case of solar power. During the day, cloud cover is an important factor. As with wind, climate variability and climate change may have a bearing on longer-term patterns of solar radiation at the surface in particular locations.

Solar array in the Mojave Desert, California

Weather can also affect the power supply in a very different way: severe weather events, such as gales and freezing rain, can cause damage to transmission lines. Forecasts can help network operators to plan mitigating actions in good time.

THE ROLE OF CLIMATE FACTORS ON BUILDING CONSTRUCTION

The effect of meteorological factors on human constructions has long been considered so much so that in the traditional architecture and new civil engineering, this issue has been accounted for according to the available facilities and information. Considering climate factors registered based on long-term climate data is essential in designing and building construction in different locations so as to assimilate with the region's climate and minimize the potential adverse effects and also to optimize the climate potentials.

Also, there has never been any study on the use of weather conditions in near future about the adjustment of civil operations especially on weather and climate disasters like flood, heavy rain and snowfall, heavy wind, etc. in any scientific papers.

Climate Factors

There are different climate factors that must be considered in civil operations, building constructions and building designs. The most significant climate factors are as follows: weather temperature, soil temperature, angle and intensity of sunlight, relative humidity, direction and wind speed, rainfall and sunlight. Climate factors cannot be reduced in these items. Rather barometric pressure and such are also considered as climate factors but they don't play a big part in designing and civil operations. Besides, the changes of some climate parameters are not high over the year so it does not make a big difference knowing about their time changes. Now, we'll be dealing with each of these factors.

Weather Temperature

Perhaps weather temperature is the most important climate factor affecting climate designing. The intended dimensions in designing various points of a building and also the material in use are determined by the maximum and the minimum temperature of the region. Therefore, the quantity and quality for constructing a building are different depending on the type of the region: tropical, cold and moderate. Glacial region would require taking special decisions on the choice of materials. In order to prevent energy dissipation in tropical and cold regions in summer and winter time, body insulation of the buildings must be considered whereas this issue might not be of priority in moderate regions.

Soil Temperature

Soil temperature and its changes are of importance over the year. The soil surface experiences the most changeability of temperature during the year which is caused by proximity with air and its changes. The more we go to the depth of the soil, the less changes of temperature we have so that in a specific depth called depth or attenuation depth temperature, annual changes of soil temperature is caused. Given the fact that building foundation lies in the soil, knowing about soil temperature, especially glacial soil and its depth, is of high significance in the selection of the materials and determining the foundation of a building. Moreover, knowing about the depth of glacial soil can be effective in the installation of gas, water pipes. The depth of glacial soil is a point where the soil does not freeze in the coldest time of the year. It is clear that in cold regions the depth of installing such equipment must be lower than the depth of glacial soil so that they become immune of any frozen state. In order to know about the depth of glacial soil in any region, one shall refer to long-term climate data provided by meteorological stations. Now that the question of constructing urban common tunnels is around, this issue's importance is doubled.

Sunshine

The rate of receiving sunshine is a function of several factors which include: latitude (angle of sunshine), the amount of cloud, and sunshine hours. The less the latitude, the less the angle of the stretch of sunshine with the vertical line to horizon and the more the receiving sunshine. Sunshine hour is irrelevant with the amount of cloud, when one increases the other decreases. In areas where the rate of receiving sunshine by the earth surface is high, temperature is high too. If the goal is to reduce the amount of sunshine entering the earth surface (this method has recently been noticed by climatologists and it is one of the strategies for reducing global warming and earth engineering), we can use reflexive mirrors.

Relative Humidity

By definition, relative humidity refers to the proportion of the existing amount of humidity in the air to the maximum amount of acceptable humidity in terms of percent. The more the rate of relative humidity, the more the possibility for the formation of water drops on physical objects on the earth surface (including buildings and other constructions such as bridges, streets,etc.). This means the acceleration of the effect of humidity on equipment and their rust chemically (corrosion of metals, oxidation of metals, etc.) and physically (freezing water and causing crack in building design. In the regions where there is more relative humidity of weather such as coastal areas and islands, designing and construction of the buildings take place according to high adverse effects of water. They must be designed and constructed in a way that physical and chemical adverse effects of water decrease to the minimum level or even zero. This need is met through the selection of water and corrosion-resist materials and equipment.

Wind Direction and Speed

Wind direction is a way from which it is blowing. Knowing about the wind direction of each region, bearing the most frequency from that direction (prevailing wind), is an important factor in setting the direction of building construction aerodynamically so that in the state of heavy winds, light buildings wouldn't be hit. In ancient times, in order to design the direction of wind wards especially in tropical regions (Yazd and Kerman), the length of wind ward vents was built in the direction of prevailing wind so that wind could be used in the best way possible to cool the building.

Wind speed is also important because in the case of high speed winds, there is the possibility for the detachment and physical damage to different parts of building especially light ones. Knowing about the mean of wind speed at the project site and seasonal and annual distribution of wind speed are important factors for strengthening against wind power. The more the mean of wind speed in the region, the more powerful the building must be.

In projects where there are several choices in terms of region, it is a good idea to pick up the one in which wind speed is lower than other regions. These regions are identified through wind pattern plans and measuring wind speed in different points of a region. If the goal is to use wind power such as establishing wind power plants, the place of the construction shall be chosen in a way that there is high-speed wind around. It is clear that, in this case, permanent high-speed wind for most of the year or even all the year is an upward.

Rain Fall

The amount of rainfall is one of the most determining factors that shall be considered in building design, especially ceiling design. In rain areas, the ceiling of buildings must be designed as gable roof so that water erosion is reduced, due damages are minimized and there would be no water left on the roof. Otherwise, adverse effects of rain and its penetration into buildings would rise. Knowing about the rate of rainfall, especially for designing structures like dams (estimation of the maximum probable rainfall) is necessary so we can determine spillway dimension, etc. moreover, in order to design surface water disposal system across cities when it rains, knowing about the maximum urban flood relevant to return period sounds like essential.

Weather Detrimental Phenomena

It refers to so high or so low rate of meteorological factors. For instance, flood or draught shows very high or low rate of rainfall. What matters about weather detrimental phenomena in civil operations is so high rates of meteorological factors. For example, concerning rainfall, its high rate which is flood or heavy showers is important and draught is unimportant. High warm and cold waves happening recently in different parts of the world due to climate change are considered to be important disasters. Tsunami receives high attention in coastal areas and islands. Dust storm makes limitations in the areas prone to wind slide. Limit values of weather and climate factors also happen in local conditions which require local studies and must be analyzed before starting the project.

Weather Forecast

It is essential to know about weather conditions of some days ahead so that you can adjust civil operations and plan doing different steps of building operation. The most important example is to know about rainfall occurrence during working days ahead. Rainfall occurrence would challenge building operations especially in primary stages of the project and can even strand it. This limitation in later stages when building constructions take place indoor is by far reduced.

One of the important uses of meteorological forecast is about dam-building workshops. Heavy rainfall occurrence in lands above the dam can cause flood in the workshop site and damage the existing equipment and vehicles and even claim human lives. Forecasting the amount of rainfall in the upper land and estimating the potential amount of the time of flood discharge, one can inform the workshops appropriately and prevent financial loss and human death.

Wind forecast can also play a big role in doing civil operations and construction especially about sky scrape buildings. Wind speed goes up with height increase and the increase in wind speed is logarithmic. Working in heavy wind conditions especially in highlands would cause damage to equipment and personnel. That is why it matters.

Non-liquid rainfall such as snow, hail, etc. brings about their own special problem. Forecasting this condition beforehand would help us adjust civil operations and timing for days ahead.

It is worth noting that weather forecast is not always true and it can be different in terms of precision and correctness according to various conditions and factors. So, it is possible for weather forecast to go wrong. In this case, ceasing the administrative operations for a while would put the project off and cause financial loss. Considering risk management conditions, the amount of this potential financial loss comparing to the loss coming from weather detrimental conditions and also human loss is negligible. Of course, taking all scientific forecast principles, the total amount of forecast would reduce to less than 10% in a period. Again, compared to relative potential material damages, the total potential material loss is lower.

PERMISSIONS

All chapters in this book are published with permission under the Creative Commons Attribution Share Alike License or equivalent. Every chapter published in this book has been scrutinized by our experts. Their significance has been extensively debated. The topics covered herein carry significant information for a comprehensive understanding. They may even be implemented as practical applications or may be referred to as a beginning point for further studies.

We would like to thank the editorial team for lending their expertise to make the book truly unique. They have played a crucial role in the development of this book. Without their invaluable contributions this book wouldn't have been possible. They have made vital efforts to compile up to date information on the varied aspects of this subject to make this book a valuable addition to the collection of many professionals and students.

This book was conceptualized with the vision of imparting up-to-date and integrated information in this field. To ensure the same, a matchless editorial board was set up. Every individual on the board went through rigorous rounds of assessment to prove their worth. After which they invested a large part of their time researching and compiling the most relevant data for our readers.

The editorial board has been involved in producing this book since its inception. They have spent rigorous hours researching and exploring the diverse topics which have resulted in the successful publishing of this book. They have passed on their knowledge of decades through this book. To expedite this challenging task, the publisher supported the team at every step. A small team of assistant editors was also appointed to further simplify the editing procedure and attain best results for the readers.

Apart from the editorial board, the designing team has also invested a significant amount of their time in understanding the subject and creating the most relevant covers. They scrutinized every image to scout for the most suitable representation of the subject and create an appropriate cover for the book.

The publishing team has been an ardent support to the editorial, designing and production team. Their endless efforts to recruit the best for this project, has resulted in the accomplishment of this book. They are a veteran in the field of academics and their pool of knowledge is as vast as their experience in printing. Their expertise and guidance has proved useful at every step. Their uncompromising quality standards have made this book an exceptional effort. Their encouragement from time to time has been an inspiration for everyone.

The publisher and the editorial board hope that this book will prove to be a valuable piece of knowledge for students, practitioners and scholars across the globe.

INDEX

www.ingramcontent.com/pod-product-compliance
Lightning Source LLC
LaVergne TN
LVHW062129200926
848484LV00007B/1789